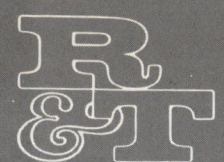

D1741348

£8.95

ROAD&TRACK

ON

VOLVO

1957-1974

Reprinted From
Road & Track Magazine

ISBN 0 948207 30 2

Published By
Brooklands Books with permission of Road & Track
Printed in Hong Kong

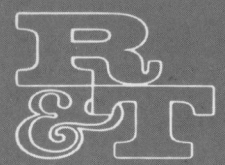

Distributed By

Road & Track
1499 Monrovia,
Newport Beach,
California 92663, U.S.A.

Brooklands Book Distribution Ltd.,
PO Box 146,
Cobham, Surrey KT11 1LG,
England

Printed in Hong Kong

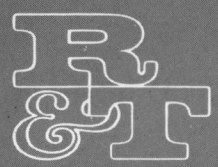

Contents

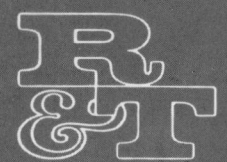

We are frequently asked for copies of out of print Road Tests and other articles that have appeared in Road & Track. To satisfy this need we are producing a series of books that will include, as nearly as possible, all the important information on one make or subject for a given period.

It is our hope that these collections of articles will give an overview that will be of value to historians, restorers and potential buyers, as well as to present owners of these automobiles.

ROAD TEST: **VOLVO PV444**

I**N** Southern California, where advertising takes on strange shapes and forms, the Volvo has recently received more high-pressure sales treatment than any other imported car. Its name has been seen on signboards, buses, streaming behind a slow biplane, and radio and television audiences have been urged to "Go, go, go in a new Volvo!" For a perfectly respectable but entirely unrevolutionary two-door family sedan, all this fancy huckstering seems a little incongruous, but it appears to have worked its peculiar magic to the extent that in some areas Volvo sales have been running second only to Volkswagen among the imports. And in whatever characteristics of modern automotive design the Volvo may be lacking, one thing is certain: for a sedan of its engine size, go, go, go it does, does, does.

Aktiebolaget Volvo of Gothenberg, Sweden, have been manufacturing automotive products since 1924, and their range includes heavy duty trucks, commercial vehicles, diesel and gasoline powerplants, and farm equipment (one small tractor uses the passenger car engine, suitably modified). The proper designation of the sedan tested here is the PV 444; now, in engineering PV stands for pressure-velocity (referring to the load-speed factor in bearings), but whatever the Swedish intent, the initials seem appropriate if only because the car has lots of "pressure" and plenty of "velocity." The 444 part is a little outmoded, since it refers to an older model which had 4 cylinders and 44 bhp; the current version sold in Europe develops 51 bhp, and the U.S. import model uses what is called the "Sports" engine which puts out 70 bhp at 5500 rpm.

Looking at the Volvo for the first time, one cannot help but be struck by the fact that the body design is of a type not built in this country since the late 'thirties. The split windshield, fat, curved rear section, squarish front end and fenders are reminiscent of such cars as the '39 Dodge (front) and '41 Ford (side and rear). So if you've got to have fins and furbelows, look no longer towards the Volvo; but if economy and performance are what count, consider this: no other sedan of under 1.5 litres we have ever tested has turned in acceleration times up to 60 mph equal to this Scandinavian import. There is nothing lavish or gaudy about the interior. Four people can ride comfortably, five with a squeeze. Seating is firm, but there is ample height (41 in. front; 37½ in. rear, seat to roof) and adequate leg room (except for rear passengers when the front seats are set far back). The front seat-backs will adjust for angle (though not conveniently, since shims must be removed), and for hardy ones who want to sleep in their car a "bed set" is available consisting of fasteners and supports wherewith the cab can be converted into a boudoir. Interior trim is plain throughout, but everything gives the appearance and feel of being solidly made and reliably put together.

Riding qualities are all on the firm side. Of course, we did use 28 psi in the tires when the factory recommends 18 & 21, but we honestly felt that the recommended pressures actually impaired the car's excellent handling qualities. Suspension is by coils all around, and as a result of their firmness there is almost no roll except when cornering extremely hard. The car steers very easily requiring a minimum of effort at any speed; steering is almost neutral with, if anything, just a touch of understeer. The only disconcerting

Unusual (these days) is the sturdy floor mounted "wobble-stick."

The sports engine turns up to 7000 rpm with 172 lb. valve springs.

Photography Poole

handling characteristic was noticed on a fast bend where even a light bump would cause the front end to jump sideways. Turns of the steering wheel required from lock to lock are 3.2.

Since the standard engine develops 51 bhp, it is easy to understand that the process of extracting 70 horses from 1414cc has not been accomplished without some penalties. The Sports engine is definitely noisy and rough compared to other 1.5 litre powerplants, and despite rubber mountings which allow the unit to rock gently at idle, there is no denying that considerable vibration is present, particularly between 50 and 60 mph in high gear. This is roughly in the 3000-3600 rpm range. The small air-cleaners on the twin SU carburetors also tend to be noisy and perhaps account for a rather peculiar but apparently harmless sound heard during deceleration.

A three-speed transmission is used in the Volvo, and its ratios are rather widely spaced, with an especially large jump from 1st to 2nd. In fact 2nd is more like the 3rd gear in most four-speed boxes. British reports (on the 51 bhp version, the 444K) comment on gear noise, but in our test car it was neither worse nor better than an average British car of comparable size and cost. There are times, how-

a sturdy import from Sweden that tops in class in performance

ever, when a synchromesh low would be appreciated because 2nd gear is rather "flat" below 20 mph.

One astounding item is the almost astronomical revolution rate of this little pushrod engine; to check valve bounce we once reached an indicated 75 mph in 2nd gear, and even allowing for speedometer error (about average), such a velocity requires 7000 rpm. Examination of the acceleration curve shows that acceleration is tremendous (for a 1.5 litre sedan) up to 60 mph. High gear, however, has been chosen for best possible top speed as indicated by the fact that the bhp peak (5500 rpm) gives 91.8 mph—by coincidence exactly what we attained on the best run. This speed is a long time in coming (nearly two miles were required), and the fall-off in pulling power when shifting from 2nd to high is graphically illustrated in our usual chart. In fact, the high gear pulling power is only 160 lbs/ton, the Tapley meter holding this figure steadily between 50 and 60 mph. An optional 5.43 "mountain ratio" is available, but we do not recommend such an extreme ratio except for short-course production sedan racing where the theoretical top speed would be about 87 mph at 6200 rpm. This may, indeed, account for much of the phenomenal success of the Volvo in West Coast competition because with such a ratio the car would get up to its top speed (out of a corner) on even a short straight in quite a hurry. During the recent contest at Pomona, the sedan race was run along with the ladies' race, and although a Porsche Spyder finished first, there, bounding along in 2nd place overall, came a Volvo, comfortably ahead of all kinds of sports machinery. Not that many owners will want to race their cars, but it's reassuring to know that the push and handling are there on tap.

In a commendable effort to bring their product esthetically up to date, the parent company introduced at the London show last fall a new 1600cc, 4-door model called the "Amazon." It is not in production yet, and if it becomes available at all, the time will probably not be until late this year or early in 1958. Also, it will have a considerably higher price than the very moderate tag on the PV 444.

While the Sports engine makes itself heard and felt to some extent, anyone who wants performance from a small, low-priced import will be highly pleased with the Volvo PV 444. Its only serious shortcoming is the lack of a "gear" between 1st and 2nd. The company seems to be aware of this, and an alternative gearbox with *five* forward speeds is being readied as an extra cost option. For those interested, the ratios will be 1.00, 1.35, 1.85, 2.80, and 4.0. And as standard equipment at no extra cost an exhaust system is furnished which at certain speeds produces a "rap" embarrassingly reminiscent of that old party standby, the Whoopee Cushion. ●

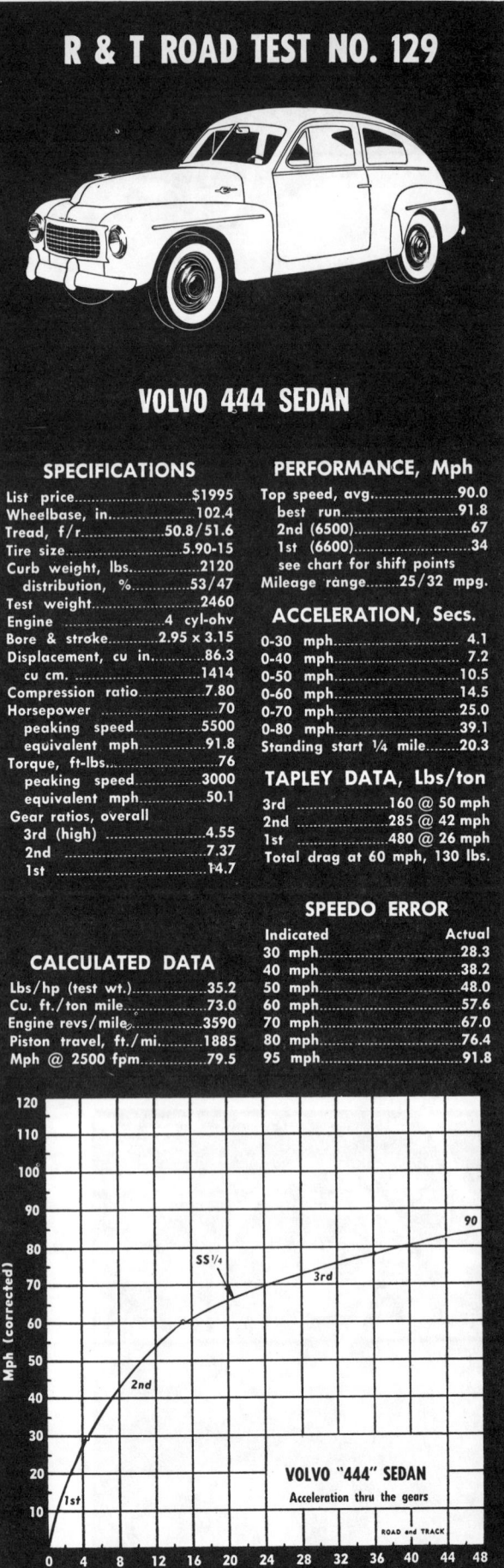

R & T ROAD TEST NO. 129

VOLVO 444 SEDAN

SPECIFICATIONS

List price	$1995
Wheelbase, in.	102.4
Tread, f/r	50.8/51.6
Tire size	5.90-15
Curb weight, lbs.	2120
distribution, %	53/47
Test weight	2460
Engine	4 cyl-ohv
Bore & stroke	2.95 x 3.15
Displacement, cu in.	86.3
cu cm.	1414
Compression ratio	7.80
Horsepower	70
peaking speed	5500
equivalent mph	91.8
Torque, ft-lbs.	76
peaking speed	3000
equivalent mph	50.1
Gear ratios, overall	
3rd (high)	4.55
2nd	7.37
1st	14.7

CALCULATED DATA

Lbs/hp (test wt.)	35.2
Cu. ft./ton mile	73.0
Engine revs/mile	3590
Piston travel, ft./mi.	1885
Mph @ 2500 fpm	79.5

PERFORMANCE, Mph

Top speed, avg.	90.0
best run	91.8
2nd (6500)	67
1st (6600)	34
see chart for shift points	
Mileage range	25/32 mpg.

ACCELERATION, Secs.

0-30 mph	4.1
0-40 mph	7.2
0-50 mph	10.5
0-60 mph	14.5
0-70 mph	25.0
0-80 mph	39.1
Standing start ¼ mile	20.3

TAPLEY DATA, Lbs/ton

3rd	160 @ 50 mph
2nd	285 @ 42 mph
1st	480 @ 26 mph
Total drag at 60 mph, 130 lbs.	

SPEEDO ERROR

Indicated	Actual
30 mph	28.3
40 mph	38.2
50 mph	48.0
60 mph	57.6
70 mph	67.0
80 mph	76.4
95 mph	91.8

VOLVO "444" SEDAN
Acceleration thru the gears

ROAD and TRACK

Often compared to a '41 Ford, the Volvo has a certain pleasing quality. Who knows, maybe the '41 Ford is coming back.

ROAD TEST VOLVO PV-444-L

ONLY A FEW MONTHS AGO, in April, we tested the 70-bhp Volvo. Now, along comes a real surprise, the same car with 85 bhp. Everyone remarks about the similarity of appearance between the Volvo and a 1941 Ford. Now we can add another Ford feature of that era, the 85 horsepower. Volvo called the 70-hp model the PV-444-K; the new 85-hp model is officially the PV-444-L.

The new engine (with a larger bore) is designated as the B-16-B, and already some sources are casting strong doubts as to the accuracy of the advertised bhp. Simply on the basis of a displacement increase from 1414 cc to 1577 cc, the power should go up from 70 to 78. But the compression ratio has been raised from 7.8 to 8.2:1, and this will add further to the output. Also, the torque peak now occurs at 3500 rpm (formerly 3000) which would indicate a camshaft change. Accordingly, we see no reason to doubt the ability of this engine to produce as claimed.

As a matter of fact, we essayed a rather extensive series of Tapley meter tests, toward the end of determining the exact rear-wheel horsepower. We were hampered by a low-speed carburetion fault, and the results were inconclusive. This much we do know: the Tapley readings of pulling power indicated more than the claimed 14.5% increase in torque.

The carburetion fault was corrected by our supplier (Ron Pearson, the invincible Volvo exponent) but even so, the cold figures show that the 0 to 30 and 0 to 40 times were not quite so good as before. This was hard to explain until we discovered that low gear has been altered slightly, from 3.23 to 3.13. The most impressive performance gain found is in high gear and above 60 mph. The improvement is shown graphically on the acceleration chart.

The average timed top speed proved to be 93.8 mph, or 3.8 mph more than in the earlier test. Such a speed is truly astonishing for a 1.6-liter sedan. A rough calculation shows that this increase in top speed would require 8 more bhp at the rear wheels. (Based on $cw = .5$ and $A = 22$ sq. ft.)

Putting all considerations of performance aside, the Volvo is still a tremendous automobile as a sturdy and practical

Chock full of machinery, the engine room shows no exterior change from the earlier 70-hp version.

Pleasing use of trim to compliment rather than to ornament the lines of the body . . .

Blue and cream plastic interior suggests a very expensive customizing job.

with 85 bhp, the sturdy Swede comes out swinging

utility sedan. When really thrashed the fuel consumption drops to 23 mpg, but normal 55/60-mph highway cruising will give 27 mpg as a best figure. It will cruise comfortably and easily at 75/80 mph, and under light throttle application the power unit is smooth and quiet. Unfortunately, the vigorous sports character of this unit becomes quite apparent when it is pushed hard. Under full throttle it seems to vibrate and becomes noticeably rough and noisy. With fond recollections of the 1931 PA Plymouth's smoothness, we fail to see why a small 4 should be quite so harsh as this one. Yet there is no question but that this is as tough a little engine as you will find anywhere, today.

Chassis-wise, the new Volvo continues with its proven unit construction. Road-rumble has been well subdued. As a matter of fact, the Volvo is not a light car (this one had a radio and weighed 50 lb more than our 70-bhp test car) and it uses heavier than normal gauge steel in many body and structural parts. The solid rear axle is located by a long rubber-insulated trailing arm on each side and uses coil springs. This and an equally well insulated front suspension of conventional design are responsible for an excellent ride, moderate roll, and generally good handling qualities.

The steering, as before, requires 3.2 turns and is light in action, with moderate understeer. Cornered really hard, there is perhaps more roll and a shade more caster return than a sports car driver would like, but a family car man (or woman) will never complain about this. At over 80 mph the steering seems to get "light" and is a little vague, but not so sensitive as to be frightening. Freedom from road shock transmission to the steering wheel is excellent.

Clutch action is unobtrusive, with no sign of slip at any time. The brakes were used fairly hard on several occasions. They, too, are eminently satisfactory. The 85-hp car has more brake lining area than the other version.

Externally, the new model can be identified by the tubular bumper guards at both ends and a new trim around the grille. The interiors are substantially unchanged, except that two-tone plastic upholstery tends to brighten things up considerably. A heater and defroster are standard equipment, but there was no opportunity to try these.

We understand that plans for producing the sports roadster and a five-speed gearbox have been completely abandoned, but with a car like this—who needs a sports car?

VOLVO 85-HP SEDAN

SPECIFICATIONS

List price	$2095
Wheelbase, in.	102.4
Tread, f/r	50.8/51.6
Tire size	5.90-15
Curb weight, lb	2170
distribution, %	53/47
Test weight	2490
Engine	4 cyl, ohv
Bore & stroke	3.125 x 3.15
Displacement, cu in.	96.2
cu cm.	1577
Compression ratio	8.20
Horsepower	85
peaking speed	5500
equivalent mph	91.8
Torque, lb-ft	87
peaking speed	3500
equivalent mph	58.5
Gear ratios, overall	
3rd (high)	4.55
2nd	7.38
1st	14.3

PERFORMANCE, Mph

Top speed, avg.	93.8
best run	95.2
2nd (6500)	67
1st (6600)	35
see chart for shift points	
Mileage range	23/29 mpg

ACCELERATION, Sec.

0-30 mph	4.3
0-40 mph	7.2
0-50 mph	10.3
0-60 mph	14.3
0-70 mph	21.0
0-80 mph	29.0
0-90 mph	44.5
Standing start 1/4 mile	19.5

TAPLEY DATA, Lb/ton

3rd	200 @ 52 mph
2nd	340 @ 44 mph
1st	540 @ 27 mph
Total drag at 60 mph, 117 lb	

SPEEDOMETER ERROR

Indicated	Actual
30 mph	28.6
40 mph	38.0
50 mph	47.9
60 mph	57.5
70 mph	67.0
80 mph	76.3
90 mph	85.5
103 mph	95.2

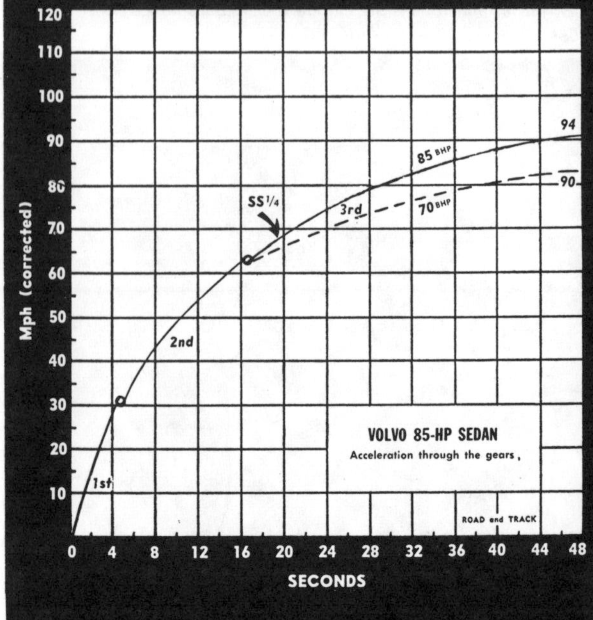

CALCULATED DATA

Lb/hp (test wt)	29.4
Cu ft/ton mile	80.4
Engine revs/mile	3590
Piston travel, ft/mile	1885
Mph @ 2500 ft/min.	79.5

VOLVO 85-HP SEDAN

Acceleration through the gears,

ROAD and TRACK

ROAD TEST 4-SPEED VOLVO

THE VOLVO has steadily, since the date of its first appearance in the U.S., continued to make friends. In the past it has been because of some fine virtues: economy, performance, sturdy construction and rugged dependability, and in spite of some drawbacks: the resemblance to a 1940 Mercury with 1948 Ford fenders (all reduced in size), and a strictly American-style, 3-speed transmission. These latter "features," of course, may appeal to enthusiasts of older-model Fords and therefore would not be considered drawbacks by them.

Well, it still looks like the aforementioned hybrid refugee from Dearborn, but it now has a wonderful 4-speed, all-synchromesh transmission. Still operated by a long, floor-mounted lever, as in past models, it nevertheless is smooth in operation and adds to both the performance and the pleasure of driving the car. The only possible criticism of the transmission might be the spacing of gear ratios. Second and 3rd are just a mite too far apart.

This fault will not bother the driver in normal city driving, but only when maximum performance is desired or, occasionally, when ascending hills that may be too steep for 3rd yet not steep enough to require the use of the much lower 2nd gear.

When maximum speed in low gear is reached and a quick shift into 2nd is made, the car continues to accelerate in a brisk fashion. But when the car is all extended in 2nd gear and a shift made to 3rd, it bogs down a little due to the wide gap between the gear ratios. Then, with an upshift from 3rd to 4th, acceleration is steady once again.

In spite of the long floor shift lever, the gears were always easily engaged and the transmission is a real tribute to its designers. A remote shift in place of the long lever would be like having egg in your beer, and it should be possible without altering the passenger compartment at all, due to the individual front seats.

There has always been a certain amount of comment, from readers as well as the distributors of the cars tested, about the figures obtained during our road tests.

The tests conducted by Road & Track's staff are road tests, not destruction tests. Each car tested is driven as though it belonged to us, and therefore it is neither thrashed nor coddled. It is given as much consideration as our own personal cars. Accordingly, acceleration and gas mileage figures obtained by us should also be obtainable by an individual owning the same make and model car.

In the case of the Volvo, a mileage check resulted in an actual average of 25.8 miles per gallon in normal city driving. If the car were driven extremely hard this mileage

Engine and accessories are easy to get at and service.

PHOTOGRAPHY: POOLE

Seat mounting leaves foot room for rear-seat passengers.

Finer mesh grille is one of few changes in appearance.

a family-type sports car that really is

would fall off, and on the open highway, or if treated with extreme care, it would, of course, improve.

No changes have been made in the chassis/body of the Volvo, and apparently none are contemplated in the immediate future, so the handling and riding qualities have not changed from previous models (R&T tests, April 1957 and September 1957). Enthusiasts will welcome the added gear, though, as it does enable the driver to select a more suitable ratio to fit each occasion.

The engine still has the roughness that was mentioned in past tests (although to a lesser degree) but is one of the most free-revving rocker-arm engines we've seen. During the test, 80 mph (indicated) was reached in 3rd gear, and the car was still picking up speed when road conditions made it necessary to slow down. This works out to an actual 75 mph, at which the engine rpm was 6500, with no valve float indicated.

The company's claims for a family sports car are obviously not without justification, and if the prospective purchaser of an economy car is satisfied with the appearance of the Volvo he would be wise to give it consideration. There is ample reason to believe he will be happy with the car and can expect, and get, a long, trouble-free life from this Swedish product.

Bumper bracing is for U.S.-style bash and smash parking.

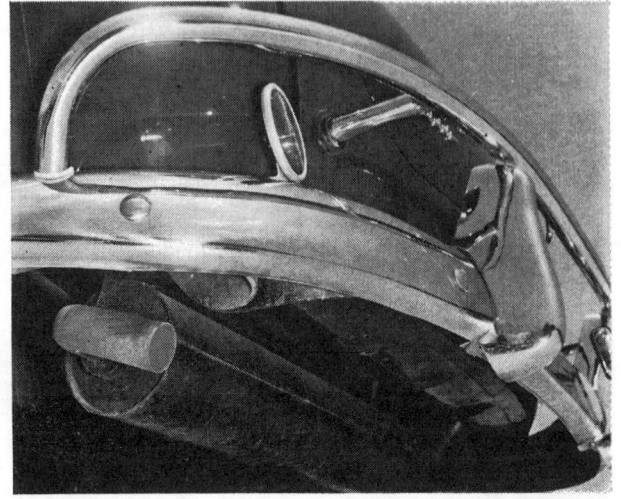

4-SPEED VOLVO

SPECIFICATIONS

List price	$2360
Curb weight	2160
Test weight	2490
distribution, %	50.5/49.5
Dimensions, length	177
width	62.2
height	61.4
Wheelbase	102.4
Tread, f and r	50.8/51.6
Tire size	5.90-15
Brake lining area	147
Steering, turns	3.2
turning circle	36
Engine type	4 cyl, ohv
Bore & stroke	3.125 x 3.15
Displacement, cu in	96.6
cc	1584
Compression ratio	8.20
Bhp @ rpm	88 @ 5500
equivalent mph	92.0
Torque, lb-ft	90 @ 3500
equivalent mph	58.5

GEAR RATIOS

O/d ( ) overall		
4th (1.00)		4.55
3rd (1.31)		5.97
2nd (2.18)		9.93
1st (3.45)		15.7

CALCULATED DATA

Lb/hp (test wt)	29.4
Cu ft/ton mile	80.4
Mph/1000 rpm (4th)	16.7
Engine revs/mile	3590
Piston travel, ft/mile	1885
Rpm @ 2500 ft/ min	4760
equivalent mph	79.5
R&T wear index	67.6

PERFORMANCE

Top speed (avg), mph	93.5
best timed run	95.0
3rd (6450)	82
2nd (6500)	50
1st (6500)	31

FUEL CONSUMPTION

Normal range, mpg	25/29

ACCELERATION

4-30 mph, sec	4.2
0-40 mph	6.8
0-50 mph	9.9
0-60 mph	13.0
0-70 mph	18.8
0-80 mph	27.0
0-90 mph	
0-100 mph	
Standing ¼ mile	19.1
speed at end, mph	71

TAPLEY DATA

4th lb/ton @ mph	195 @ 55
3rd	245 @ 50
2nd	380 @ 37
1st	540 @ 22
Total drag at 60 mph, lb	117

SPEEDOMETER ERROR

30 mph	actual 29.2
40 mph	38.3
50 mph	47.4
60 mph	56.2
70 mph	65.2
80 mph	74.8
90 mph	84.5
102 mph	95.0

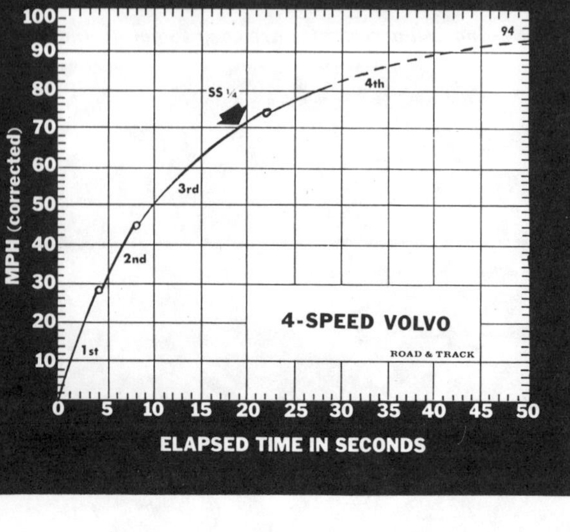

4-SPEED VOLVO

ROAD & TRACK

MPH (corrected) vs ELAPSED TIME IN SECONDS

ROAD TEST VOLVO 122-S

A newer and more attractive product of superb Swedish engineering

THE VOLVO 122-S, called the Amazon in Europe, has been built and sold in Sweden for over two years, but it is new to the U.S. market.

And a refreshing new car it is, too: pleasant looking, easy (and fun) to drive, economical and durable in the extreme. It is also refreshing to find a company that actually does something to make its product safe for the occupants, and does it without making asinine statements that the public won't buy safety.

The safety-conscious Swedes have inaugurated many features we've long advocated for cars and would like to see incorporated, in some form, in all passenger vehicles. Here we have the padded instrument panel, a dished steering wheel that is attached to a column built to collapse under pressure, and a plastic package shelf on the passenger's side which folds under impact. The sun visors are of thick foam rubber construction, and seat belts (diagonal straps that extend from the floor in the center across the passenger to the door post) are available on order.

You can, of course, put safety belts in any car, and many vehicles have padded instrument panels to protect the passenger, but most manufacturers then install a row of projecting knobs, handles or switches below the padding which nullifies its effectiveness. The only deterrents to the effectiveness of the Volvo's padded panel were the two radio control knobs.

The newest import is a handsome car in a reserved way, with no evident ostentation or gaudiness. Its excellent over-all proportions carry it well past the average medium-sized sedan in appearance. From some angles, notably the side and rear, it resembles the smaller Simca Aronde (nothing wrong with that). It is quite easily lost in the traffic shuffle, not being an outstanding example of something new in the styling department. This, of course, will keep the car from getting old as quickly as some other contemporary vehicles, so it can mean money for its owner at trade-in time.

A close examination of the car, along with many miles behind the wheel, brought favorable comments from every tester and rider. Design, construction and general quality are obviously excellent, and there is a pervasive feeling of durability. Yet there is no indication of the luxury that we expected of a 4-cyl sedan in this price bracket.

With 3 less horsepower, the same gear ratios, and over-all size and frontal area similar to those of the PV-544, the performance would be expected to be about the same on both models, and so it is. The 122-S weighs

Under the hood, a fine 4-cyl, ohv engine and a fine heater, so necessary for Sweden's cold climate.

COLOR PHOTO BY RAY HALIN

PHOTOS BY POOLE

11

The pleasing proportions are evident in this side view of the 122-S.

65 lb more, and acceleration is reduced proportionately. It is our feeling that the last Volvo we tested (the PV-444, October 1958) was in slightly better tune. If so, the difference in performance was a little more than would be indicated by the two cars' specifications:

	PV-444	122-S
Weight	2160 lb	2225 lb
BHP	85 @ 5500	88 @ 5500
Torque	90 @ 3500	90 @ 3500
Gear ratio (over all)	4.55:1	4.55:1
0–30 mph	4.2 sec	4.7 sec
0–60 mph	13.0 sec	16.2 sec
Top speed	93.5 mph	91.9 mph

Road & Track has not tested a 159 PV-544 because of its similarity to the 444, but the above figures can be assumed to be accurate for this year's model.

The 1600-cc, 4-cyl engine is extremely flexible and, as we've said before, one of the most free-revving rocker-arm engines we've seen. It also seemed to run smoother than other Volvo engines.

Even low gear is synchronized in the 4-speed transmission. Still controlled by a long floor-mounted lever, as in more familiar Volvo models, it operated magnificently every time. There is still too much gap between the ratios of 2nd and 3rd, but the bright side of this is that the car starts easily in 2nd gear. Those who shift more traditionally will find that the shift lever has so strong a spring that it wants to go directly from low gear to 4th instead of 2nd. Yet the engine is so amenable that it pulls calmly (albeit with little power) from about 15 mph on up.

Both the sturdily vinyl-covered seats and the suspension of the 122-S appear at first to be a little too stiff for comfort. Longer excursions in the car emphasize the wisdom of fairly firm seat cushions (ultra-soft seats are fine on a sofa, but not in a car) but also point up the need for a more bucket-like treatment of the seats themselves; the backs are fine. The suspension grows to feel a little softer. Bad bumps, however, catch it napping.

Foot room for both front- and rear-seat passengers is good, but the knee room of rear-seat passengers is somewhat limited when the front seats are in a position from the center of travel on back. The front-seat adjustment handle itself is the biggest single annoyance in the entire car. It seems mechanically sound, but it's the poorest we've seen for safety. We strongly advise purchasers of the car to exercise extreme caution when adjusting the seat (especially rearward), lest their cut or pinched fingers require medical aid. The rear door window handles are difficult to use when the front seats are back.

The emergency brake is handily located on the driver's left and a clever protective ring prevents its accidental release. Several times during the test we were moved to wonder why they didn't mount it on the driver's right, between the front seats.

Visibility is excellent, due to the well placed window areas (not overly large) and the high seating position. After several years in cars with low builds, the seating

A pleasant but unexciting interior is functional.

The neat luggage compartment is easily accessible . . .

Double grille treatment is different and very well done.

position of a Volvo reminds us of that in a pickup truck. The upright seating, more practical than it is stylish, helps to get comfortable room for 4 passengers in the 102-in. wheelbase. The attractive appearance of the car is even more impressive in view of the excellent head room.

The unit body/frame construction, in addition to being a safer and stronger design than the separate frame and body, offers a very solid-feeling, rattle-free ride. Flinging the car around mountain curves, bumping along over secondary roads, crawling through city traffic, and cruising at 65–70 mph on the highway are all done with the greatest of ease. Corners can be taken with gusto, though with considerable body roll and squealing of tires (added pressure is advisable for faster driving). The ZF steering is precise and gives a good feel of the road to the driver. It seems slower than its 3.2 turns lock to lock.

Like the PV-444 and 544, the 122-S has coil springs and telescopic dampers all around with forged wishbones in front, coupled with a large-diameter anti-roll bar, and trailing arms at the rear with lateral axle location by Panhard rod.

The brakes are British Girling drum type, with 2 leading shoes in front. No brake fade was noticed during our tests, but considerable odor was evident after hard usage.

It has been claimed by many observers that the 122-S body is the same as that of the 1953 Willys sedan. Volvo supposedly purchased the Willys dies at the termination of production of the Willys car, but a comparison of the two makes reveals little or no resemblance. 🜔

. . . Under the large, heavily ornamented trunk lid.

ROAD & TRACK ROAD TEST 215

VOLVO 122-S

SPECIFICATIONS

List price	$2895
Curb weight	2390
Test weight	2715
distribution, %	52.3/47.7
Dimensions, length	173
width	63.5
height	59.2
Wheelbase	102.4
Tread, f and r	51.7
Tire size	5.90-15
Brake lining area	165
Steering, turns	3.2
turning circle, ft	33
Engine type	4 cyl, ohv
Bore & stroke	3.125 x 3.15
Displacement, cu. in.	96.6
cc	1586
Compression ratio	8.20
Bhp @ rpm	85 @ 5500
equivalent mph	92.0
Torque, lb-ft	87 @ 3500
equivalent mph	58.5

GEAR RATIOS

O/d (n.a.), over all		4.55
4th	(1.00)	4.55
3rd	(1.31)	5.97
2nd	(2.18)	9.93
1st	(3.45)	15.7

CALCULATED DATA

Lb/hp (test wt)	32.1
Cu ft/ton mile	72.6
Mph/1000 rpm (4th)	16.7
Engine revs/mile	3590
Piston travel, ft/mile	1885
Rpm @ 2500 ft/min	4760
equivalent mph	79.6
R&T wear index	67.6

PERFORMANCE

Top speed (4th), mph	91.9
best timed run	92.9
3rd (6450)	82
2nd (6500)	50
1st (6500)	31

FUEL CONSUMPTION

Normal range, mpg	24/27

ACCELERATION

0-30 mph, sec	4.7
0-40 mph	7.4
0-50 mph	11.4
0-60 mph	16.2
0-70 mph	22.8
0-80 mph	32.5
0-90 mph	
0-100 mph	
Standing ¼ mile	20
speed at end, mph	66

TAPLEY DATA

4th, lb/ton @ mph	175 @ 55
3rd	230 @ 50
2nd	360 @ 37
1st	450 @ 22
Total drag at 60 mph, lb	114

SPEEDOMETER ERROR

30 mph	actual 31.5
40 mph	41.0
50 mph	50.5
60 mph	58.9
70 mph	67.2
80 mph	77.6
90 mph	
100 mph	

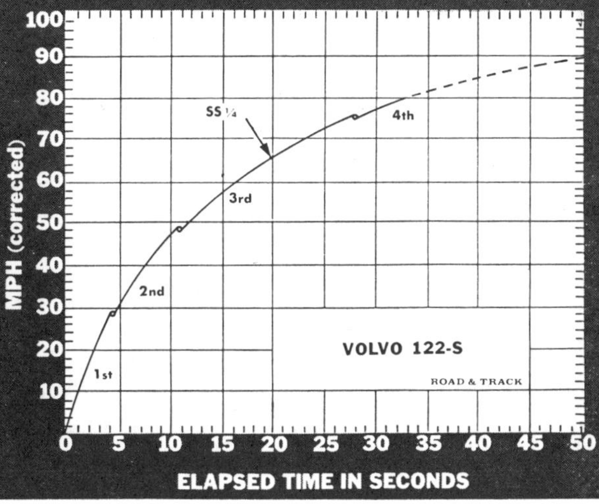

VOLVO 122-S

ROAD & TRACK

MPH (corrected) / ELAPSED TIME IN SECONDS

VOLVO P-1800

THE VOLVO P-1800 coupe made its first public appearance at the Brussels Auto Salon in January, and its first U.S. appearance at the New York International Auto Show in April.

A very pretty car, it uses many of the standard components from existing Volvo models, but a new engine has been designed specifically for the coupe. This engine is an inline ohv 4-cyl unit displacing 1.78 liters (84.1 x 80 mm bore and stroke) and producing 100 SAE horsepower at 5500 rpm.

The gearbox is based on the 4-speed unit installed in the 122-S and PV-544, but it has been redesigned for the coupe. An electrically operated overdrive is listed as an optional extra.

The brakes are servo-assisted hydraulic, with discs on the front wheels and drums at the rear; a practice that may be applied to U.S. cars in the near future.

Principal dimensions of the coupe are: wheelbase 96.5 in.; tread, both front and rear, 52; overall length 173; width 67; height 51 and ground clearance 6.3 in.

Although the body was designed in Italy, it will be produced in England by Pressed Steel, Ltd. The car will also be assembled in England, as all the available assembly capacity at Volvo, in Sweden, is being utilized by existing Volvo production.

Series production is not scheduled until September, and sometime early in 1961 production is expected to be in the neighborhood of 100 cars per week. Delivery in the U.S. has been tentatively promised for late fall of this year and no actual price has been quoted.

VOLVO P-1800

Sports car enthusiasts look closely, there may be a Fjord in your future

AB VOLVO, which manufactures a line of quality, low-priced automobiles in Gothenburg, Sweden, has been exporting cars to the U.S. for only a short time but has already established quite a good reputation with its relatively high performance economy sedans. These sedans have provided a sort of bridge between the true sports car and the utility car for many enthusiasts with growing families and for that, and other reasons, the sale of Volvo automobiles in this country has been brisk. And brisk they should be, for there are no other cars available that reach into exactly that segment of the market.

Now, Volvo is preparing to enter another specialized market—that of the medium-priced, fast Grand Touring class sports car. This is, at least for the present, an area in which not many cars are available—at least not for the under-$4000 price of the new P-1800 Volvo GT coupe. (Porsche, Alfa Giulietta, and Facellia will be the

P-1800's most obvious competition in this market.)

Volvo's P-1800 is a sports car of a type we can expect to become extremely popular. It is nothing like the traditional wind-in-the-face sports car of years past, which was good sport and all that, but had a tendency to wear on one at times. The 1800 is, rather, a very civilized touring car for people who want to travel rapidly in style, a Gran Turismo car of the type already much in the news these days—but at a price that many people who cannot afford a Ferrari or Aston Martin will be able to pay.

In order to test this new car we flew to Sweden where we were given the final-stage prototype of the P-1800 to drive, anywhere, at any speed, and in just about any fashion that should happen to strike our fancy. So, in addition to the usual banging about on the public roads—and in Sweden there are no speed limits—we were also given permission to try the car on Volvo's proving ground,

15

where there is a huge skid-pad along with the usual turns and straight stretches. The weather was quite bad—and that was good, because it gave us an opportunity to try the car under conditions that were likely to reveal any bad handling traits. As it turned out, the car never gave us a bad moment, even though the roads were very wet through most of the time that we were driving the car.

From the driver's seat it is a bit hard to tell, except for the comprehensive instrumentation, that you are not sitting in an American car, for the interior is roomy and done in the rather futuristic fashion that characterizes the U.S.-built car. The seat is comfortable and is placed so that one has a good view through the various windows and out over the steering wheel and instrument cluster. The pedals are widely enough separated to allow us big-footed Americans to drive the car without difficulty and all of the various controls are conveniently located.

When driving the car, one is immediately impressed with the fact that it has been designed mostly for tractability, rather than racing, for it is possible to plug along in top gear at just about any speed. The engine has a long, flat torque curve and there is no thudding or thumping; even if one bangs the throttle wide open at anything over idle speed the accelerating characteristics of the car have a sort of electric-motor feeling. Just apply the accelerator and the car pulls smoothly, if not too vigorously, up toward its top speed of just over 100 mph. We might mention that Volvo's development engineers have gotten several runs with this pre-production prototype that average out to about 105 mph, with a best run of 107. They expressed the opinion that after the kinks are smoothed out, the version that reaches the market will be capable of about 110—enough to satisfy most people. Actually, this car is not designed for short bursts of speed of over 100 mph, but rather for the long stretches at about 95-100. It will, in the overdrive version, run almost indefinitely at its top speed, but not too many people will want to drive that fast for any distance.

A part of the charm of this machine can be attributed to the inclusion of the 5th gear—actually a Laycock de Normanville overdrive with a ratio of 0.756. This unit adds $145 and 33 lb, both items included in our data panel. Acceleration is good even in overdrive (3.44:1), particularly from 60 mph and up where the Tapley meter hit 150 lb/ton. However, for those who do not want overdrive, the P-1800 will be supplied with a 4.1 rear axle ratio. This will reduce the acceleration ability slightly, but the transmission ratios are well chosen—and if properly used the deficiency will be very hard to detect.

Since the calculated data in the panel summary are based on an overdrive-equipped car, it may be pertinent to point out what happens when each of the three alternative ratios is used.

		TEST CAR	OPTIONAL
Axle (effective)	4.56	3.44 (o/d)	4.10
Engine revs/mile	3600	2720	3240
Mph at 6000 rpm	100	132.5	111
Cu ft/ton mile	80.7	61.0	72.7
Wear Index	68.0	38.9	55.0

As the engine develops its peak horsepower at 5500 rpm, it seems likely that a car without overdrive (i.e., the 4.10 axle ratio) will be the fastest of the three choices shown because the 132.5 mph figure (for overdrive) is purely theoretical and would require at least 50 more bhp than is available.

Of all the components of the P-1800, easily the most interesting is the engine, which embodies nothing new or exciting except that it has been designed and is being built with remarkable care. The single item that attracts the most attention is the fact that the crankshaft is carried in 5 main bearings, 2 more than the usual number. And is this important? That question can be answered much better after a couple of years or so have passed and the private owner, always the person most likely to discover design flaws, has had *his* chance at the engine. We personally consider this engine to be quite unbreakable—at least so long as it has plenty of water and oil—because we were given almost complete access to the development records (as many as we requested) and we know that the reliability test consisted of 500 continuous hours on the engine dyno pulling full-throttle and full load . . . in other words, 5500 rpm for 500 hours non-stop, and pumping out 100 bhp all the while. As one Volvo engineer put it, "The driver will tire long before the engine does." The crankshaft is a great, stout-looking thing with a goodly amount of overlap between journals, and the journals themselves are induction hardened and run in fatigue-resistant copper-lead bearings. Needless to say, the bottom end appears to be capable of withstanding far more than the present 100 bhp. The main bearing caps are a sort of *piece de resistance,* as they are also uncommonly large and have the distinction of being supported from the sides as well as being just pulled up into place by bolts. The

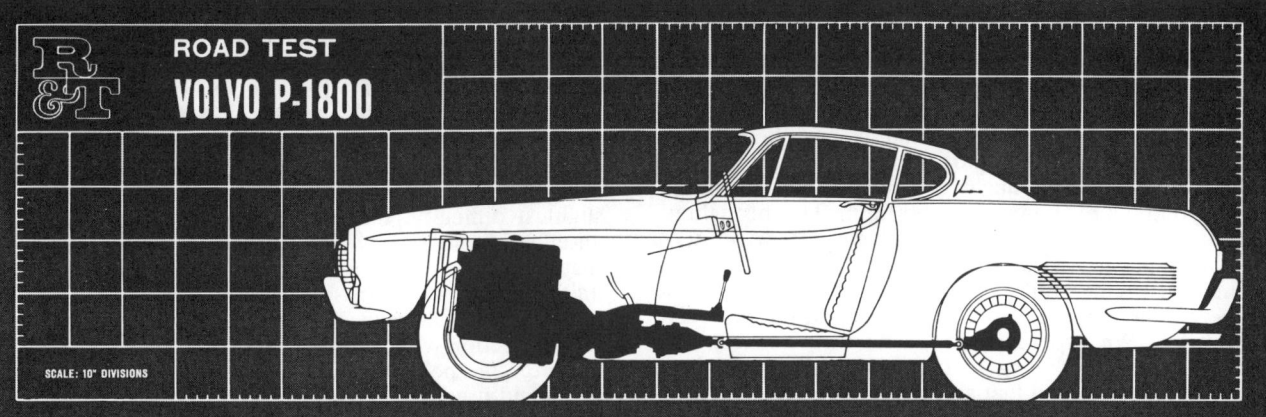

ROAD TEST
VOLVO P-1800

SCALE: 10" DIVISIONS

DIMENSIONS

Wheelbase, in	96.5
Tread, f and r	52.0
Over-all length, in	173
width	67.0
height	51.0
equivalent vol, cu ft	342
Frontal area, sq ft	19.0
Ground clearance, in	5.3
Steering ratio, o/a	15.5
turns, lock to lock	3.2
turning circle, ft	30
Hip room, front	53.5
Hip room, rear	50.5
Pedal to seat back	39.0
Floor to ground	10.8

CALCULATED DATA

Lb/hp (test wt)	28.0
Cu ft/ton mile	61.0
Mph/1000 rpm (o/d)	22.1
Engine revs/mile	2720
Piston travel, ft/mile	1430
Rpm @ 2500 ft/min	4760
equivalent mph	105
R&T wear index	38.9

SPECIFICATIONS

List price	$3940
Curb weight, lb	2500
Test weight	2800
distribution, %	51/49
Tire size	5.90–15
Brake lining area	97.7
Engine type	4 cyl, ohv
Bore & stroke	3.31 x 3.15
Displacement, cc	1780
cu in	108.5
Compression ratio	9.5
Bhp @ rpm	100 @ 5500
equivalent mph	121.5
Torque, lb-ft	108 @ 4000
equivalent mph	88.3

GEAR RATIOS

O/d (.756)	3.44
4th (1.00)	4.56
3rd (1.36)	6.20
2nd (1.99)	9.09

SPEEDOMETER ERROR

30 mph	actual, 29.5
60 mph	58.2

PERFORMANCE

Top speed (o/d), mph	105
4th (6000)	100
3rd (5950)	73
2nd (6000)	50
1st (6000)	32

FUEL CONSUMPTION

Normal range, mpg	22/26

ACCELERATION

0-30 mph, sec	3.6
0-40	5.6
0-50	9.0
0-60	12.4
0-70	16.9
0-80	23.5
0-100	
Standing ¼ mile (est)	18.0
speed at end	72

TAPLEY DATA

4th, lb/ton @ mph	200 @ 65
3rd	280 @ 48
2nd	400 @ 35
Total drag at 60 mph, lb	147

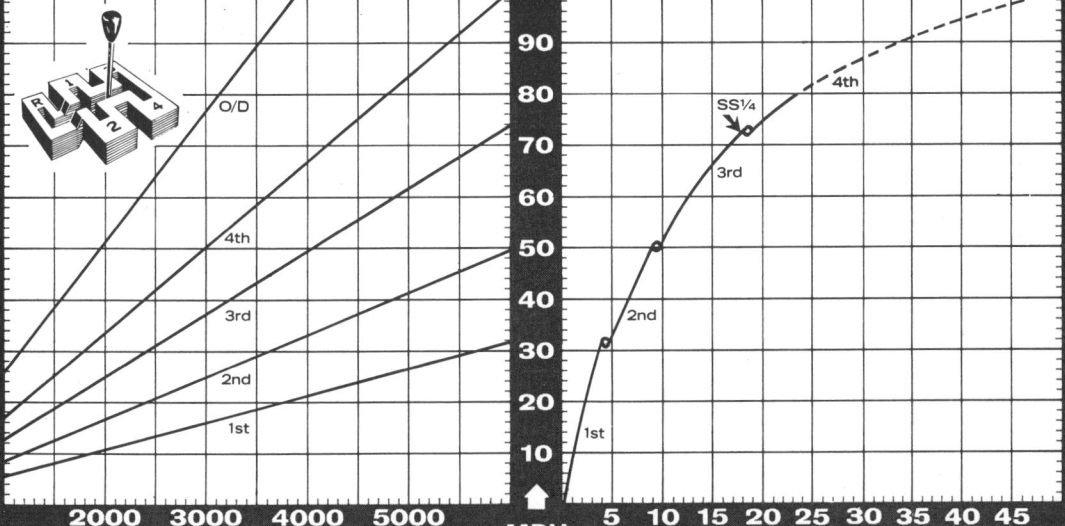

ENGINE SPEED IN GEARS

O/D
4th
3rd
2nd
1st

ENGINE SPEED IN RPM

ACCELERATION & COASTING

SS¼
4th
3rd
2nd
1st

MPH

ELAPSED TIME IN SECONDS

main bearing cap slips into a 3-sided groove in the crankcase and, as it is a tight slip-fit on each side, it looks like a rather expensive system to us, but the people at Volvo didn't seem to think that it was at all out of the ordinary.

The complete engine is a really amazing thing that has potentialities far beyond what is now being used. Once, during the driving part of the test, we expressed some concern that we might get over-exuberant and get well beyond the recommended 6000 rpm. This brought on a most unusual demonstration; Volvo's test driver took the wheel, blasted off down the road in first gear and just wound the engine until the valves began to float, then he feathered the throttle and held the engine at the valve float point. We proceeded for some distance with this ghastly clattering ringing in our ears until it had been properly demonstrated that an occasional accidental over-revving of the engine was not going to break anything. Later, we were also informed that if you persevere it is possible to get past the point of valve-float (a phenomenon caused by a resonating of the valve springs) and then the engine will turn right on up until a piston breaks, or some other type of mechanical disaster occurs. Naturally, Volvo does not recommend that the private owner try any of this unless, as Mr. Larborn, who is chief engineer in Volvo's test laboratory, says, "The private owner is prepared to underwrite the possible costs of the experiment."

The transmission, like the engine, is not particularly unusual in concept, but does its job in a most satisfactory manner. It is constructed in a proper, straightforward fashion and feels as though it would be nearly impossible to break. The synchromesh is unbeatable, no matter how fast one moves the lever, and the ratios are perfectly spaced. All four gears are synchromesh.

Brakes are probably as important as any other factor on any car, but in the case of the GT car, which is intended for high speed use on public roads, the braking system is of much more than average importance. Of course, many of the individual states in America have speed laws that severely limit the vigor with which one may drive, but in others the laws are less restrictive and one may proceed at any velocity that falls within the vague limits of "reasonable and proper." Moreover, in at least one state, and in much of the rest of the world, there are hardly any limits beyond those imposed by the nature of the vehicle in question and it is in such areas that the GT car is at its best. Under such conditions the ability to stop without any weaving or wheel locking is a prerequisite for fast cruising, and the P-1800 was designed with this in mind.

Disc brakes are used at the front and drum brakes at the rear, which is an odd mixture on the surface of things, but quite logical if one considers all of the facts. Actually, the best reason for the mixing of drums and discs is somewhat negative but true nonetheless; there is simply no point in going to the expense of disc brakes at the rear since only about a third of the braking effort is carried there. Even the least inspired of drum-type brakes will carry the effort of stopping the rear wheels without the slightest difficulty. Further, it is much easier to provide an effective emergency brake if drums are used and, last, the drum-type brakes are lighter than available disc-type units. Volvo engineers picked the Girling calipers for their car's front brakes and use these British-made units in conjunction with a disc of their own design and manufacture. One point that we noted with some interest was the shield used on the side of the disc that faces inboard. We asked about this and were told that without the shield, the pad on the inboard side of the disc was subject to vastly greater rates of wear than the pad on the outboard side, which is shrouded by the deeply dished wheel. This wheel uses a rim 4.5 in. wide to conform to the particular requirements of the tires. The tires are "low-profile" Pirelli Cinturatos, which have a special reinforced tread and are especially well suited to sustained high-speed operation. These tires really do give an outstanding grip on the road under all conditions and, we feel, contribute materially to the car's stability.

The suspension is taken directly from the parts bins that supply Volvo's 122-S, but springs of a slightly altered rate are used in the P-1800 and the Panhard rod at the rear has a revised location which provides an altered roll center. All of the front suspension and steering are carried on the front suspension member, which bolts into place rather than being an integral part of the frame.

The rear suspension is a study in refined orthodoxy, with a conventional solid rear axle. The springing medium, like the front, is coil and the axle is positioned by a pair of non-parallel rubber-bushed links on each side. Lateral movement is restricted by a Panhard rod. Because the rear axle is so firmly positioned, no rear-end "steering" is evident and the car felt secure under all conditions.

Little can be said about the body that is not apparent in pictures of the car—except that it looks somewhat larger "in the flesh." As a structure it benefits considerably from Volvo's long experience with unitized construction and from a most rigorous testing program. It is a design that typifies unit-structure practice, having a stressed floorpan and inner paneling to which the roof and outer panels add strength. And, of course, there are the usual small frame-like extensions of box-section structure that carry the suspension links and sub-frames.

The car, in its final production form, will not be a fire-breathing racing machine, but it will be absolutely dry and warm, and an altogether civilized and comfortable conveyance in which to travel about. More important, it remains comfortable right up to its top speed of just over 100 mph with very little wind or engine noise. The P-1800 is one sports car in which a good quality radio will not be wasted. Further, for the driver with sporting propensities, the handling of the car adds even more allure to the overall package. Our experiences with it, on both wet and dry surfaces, demonstrated that the P-1800 has stability, adhesion and agility, and can be drifted with *brio* without causing the driver any anxious moments.

On top of all its other virtues, the P-1800 offers the advantage of adequate luggage space, which, as many enthusiasts know, is sadly lacking in some cars they would like to own. The fact that the P-1800 is *new* is also an advantage sometimes overlooked; there are still those who want the first of anything.

VOLVO P-1800

A TECHNICAL ANALYSIS   BY GORDON H. JENNINGS

THE PATH TO SUCCESS in automobile design is not, contrary to widely held opinion, in the application of many radical technical innovations. Although these are a part of "progress" and make very good reading when broadly outlined in a sales brochure, such features all too often prove troublesome when translated into reality and placed in the hands of the general public.

At the other end of the scale we find the very dull designs that have become great popular favorites because of their reliability. Reliability in this instance is achieved by the strict avoidance of anything new, which allows the fullest application of lessons learned in development work done in the years past. Cars of this type do not do anything particularly well, but they can be depended upon to deliver their mediocre best at any time—and that is, for many of us, preferable to the exciting but uncertain kind of automobile.

Between the two extremes, ultra-radical and arch-conservative, are to be found the really good designs; designs incorporating all of the things learned so painfully in the past, but altered to take advantage of the over-all advance in technology. Naturally, this sort of philosophy would, if carried to its logical extreme, put an end to all progress. The extreme is, however, a long way off and there is plenty of room for improved automobiles within the existing, proven area of design.

A prime example of what we would term, "the refined end-product of established practice" is the "new" Volvo P-1800 sports/touring coupe, which incorporates only features of proven worth. All of them are rather old-hat, if you care to look at it that way, but all are long established as being worthwhile. Not a single radical idea, yet the design conveys a feeling of real newness because so many of the best features recently developed by the industry as a whole have been included. It may not be progress, but it promises to be an eminently satisfactory approach for today's customers.

ENGINE

The P-1800 engine, while quite undistinguished in general specification (apart from the five mainbearing crankshaft) is nonetheless an outstanding piece of work. It is nothing exotic, but is designed with such lavish attention to small detail that it can hardly be anything but successful—even in the hands of the general public: the acid test in all such matters.

Starting with the foundations of the P-1800 engine, we find a combination of crankcase and crankshaft that should allow many successive increases in engine power. Perhaps it is a bit premature to be speaking of increases now, before the car is even released to the public, but these things *are* important. The crankcase is laid out around very generous dimensions, providing plenty of room for the engine's 5 main bearings and incorporating an extensive system of reinforcing webs. The crankcase skirt has been extended down well below the crankshaft centerline—not for strength, we were told, but so that really effective, ring-type oil seals could be provided. If the crankcase had been terminated at the shaft centerline, it would have been quite difficult to avoid the semi-circular half-sections of seals that are so prone to leaking. With the system Volvo uses, the only oil escaping is that going out the crankcase breather-pipe in the form of vapor.

The P-1800's crankshaft is a real match for the crankcase in over-all massive strength. The over-square bore and stroke of 3.31 x 3.15, plus the uncommon amount of water space provided between bores, leaves a lot of space for large bearings and crank-webs. There is, moreover, a very worthwhile amount of overlap between the mainbearing and crankpin diameters. The shaft itself is a steel forging and, in addition to the usual machining and straightening operations, it has all journal surfaces induction hardened. These journals are 2.50 in. in diameter and run in bearings 1.093 in. wide—sizes more than adequate for any forseeable, or even conceivable, increase in power. Chevrolet's formidable 283 cu. in. V-8, for instance, has mainbearing dimensions of 2.30 and .762 for the journal diameter and bearing length. From this comparison, one can readily see that the P-1800 engine's main bearings are very lightly loaded, relative to their size, and should provide a long service life.

The connecting rod bearings are just as wide as the mains, run on throws 2.25 in. in diameter and are, as was the case with the mains, larger than those used in the undeniably reliable Chevrolet V-8. The connecting rods themselves are 5.72 in., measured center to center, and appear to be strong enough for anything.

The train of reciprocating parts is completed with the aluminum pistons, which have a cast-in steel strut to control the expansion of the solid piston skirt. The piston carries three rings, two compression, one for oil control, both above the wrist-pin hole. The slotted oil control ring scrapes oil from the bores and splashes it into a pair of drilled cups in the upper end of the connecting rod, where it lubricates the bronze bushing in the small end of the rod. The wristpin is held in the piston by small spring-clips and is offset slightly from the bore center to reduce piston slap when cold.

Above the crankcase, in the actual cylinderblock area, we find a layout that is—at least among cars currently produced—exclusively Volvo's. The cylinders are com-

Oil cooler fits between engine block and filter.

pletely surrounded by water space and no forced flow is provided. The cooling water begins to move only when sufficient heat is present to start a thermosyphon action— this system providing a very rapid warm-up of the cylinder walls, which reduces the bore wear rate considerably. And, as the cylinders should be relatively warm at all times, and since the thermosyphon effect produces enough circulation of water to keep the cylinders at the right temperature, there is much to recommend the Volvo system. It should also be noted that because of the exceptional amount of water space between bores, and because of the sturdy lower-end assembly, this engine could be considerably enlarged. With a little stretching here and there, the P-1800 could become the P-2000, 2500, and conceivably even 3000. However, the engine in its present form is quite adequate for the job and it should be uncommonly reliable at the existing size and power output.

Cross section drawings indicate remarkably rugged engine design.

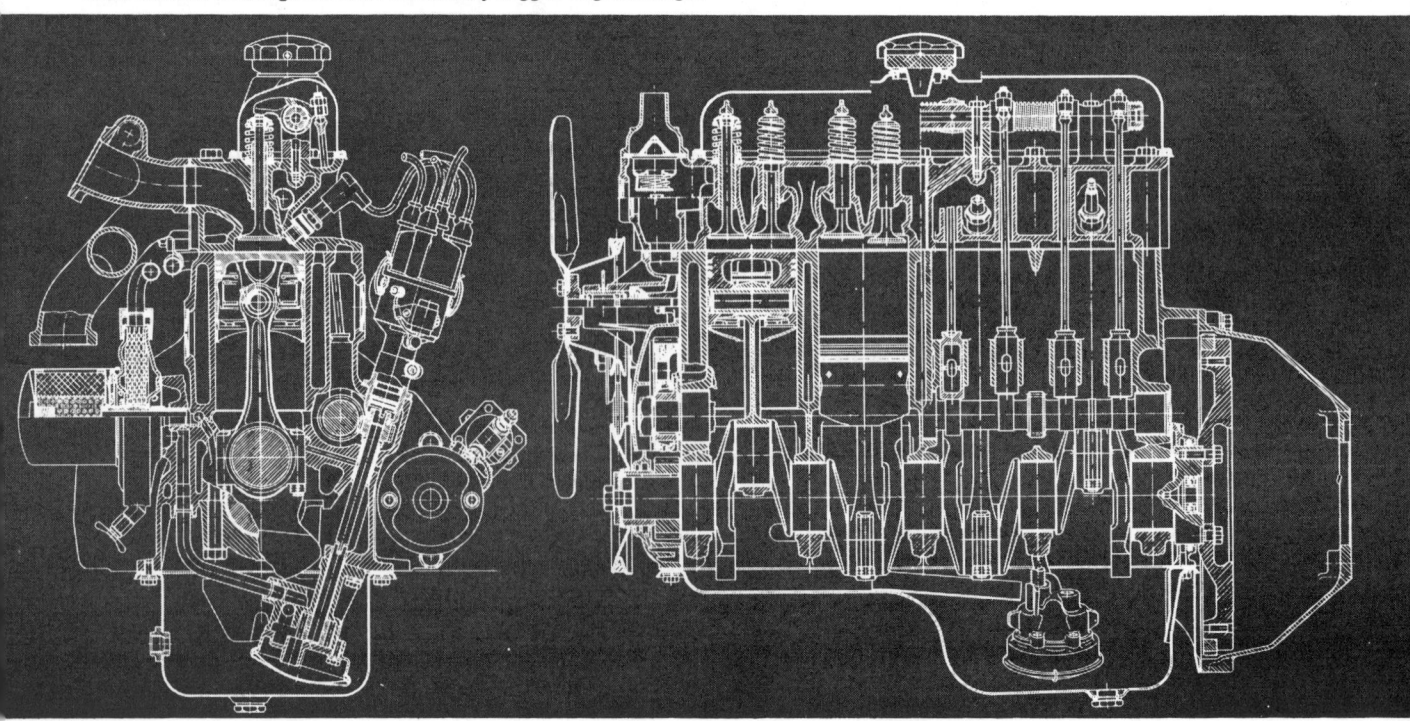

Detail photos of cylinder head, rocker-arm shaft, connecting rod, piston and valves shows nothing new or startling, but does show significant attention to design.

An additional refinement in the engine's lower end is the careful handling of the oil supply. Not only is a full-flow oil filter placed in the system to remove gritty impurities, but there is available an oil cooler. Actually, this "cooler" is a heat exchanger; it fits underneath the oil filter case and instead of radiating heat from the oil into the air, it transfers heat between the oil supply and the cooling water. Thus, when the engine is started, the water—which normally warms up faster than the oil—is used to bring the oil up to proper temperature. Then, after the normal operating temperature is reached, the heat moves in the direction required. If the car is just puttering along, the water helps keep the oil up to temperature and, if the car is being pressed hard, the water helps to keep the oil supply cool.

The cylinder head, which is—like the block—made of cast iron, bears the mark of a lot of careful planning. The valves are just about as large as could be accommodated without enlarging the bore, intakes are 1.59 in. and the exhausts are 1.37 in. in diameter. An item of particular interest is the fact that the area immediately above the valve head, in the port, is machined to shape. We were told that this is the only way in which a good, consistent port shape could be assured. Casting was not precise enough to suit the requirements.

Cooling water in the cylinder head is circulated by pump—and distributed most carefully. From the pump, water travels down the length of the head in a tubular gallery and is released from holes in the gallery situated at the exhaust valve seats. After cooling the exhaust valve seats—the most critical of all hot spots in most engines—the water moves forward, cooling the remainder of the cylinder head, to the outlet leading to the radiator.

Although the earlier Volvo passenger car engine had siamesed intake ports, the P-1800 has an "eight-port" cylinder head, with individual ports leading to each valve. These ports and the manifolding are quite well done, so pumping losses should be low. Steel spring-rings spigot into intake ports and manifold to assure that there will be no misalignment. Both intake and exhaust ports emerge from the same side of the head, but there is no provision for exhaust heat to the induction system. However, with such a short, straight induction tract leading to the ports from the two S.U. carburetors, there is probably no need for heat beyond that which is radiated from the exhaust manifold. The S.U. carburetors, incidentally, have 1¾ in. throats and are to be fitted with air cleaners that will eliminate the drawn-out "gasp" that characterized the previous Volvos.

The valve gear layout is very conventional, the valves being operated by a camshaft mounted in the block, through radius-face tappets, one-piece forged pushrods and rockers. The rockers increase the lift given by the cams at a ratio of 1.52 and the lift, measured at the valve, is .203 in. Single springs close the valves and where these press against the cylinder head there are neoprene washers which tend to dampen out some of the minor resonances. Spring resonance, when it does occur, produces a very audible valve-crash, but this phenomenon only takes place at speeds well above 6000 rpm. It was demonstrated to us, by Volvo engineers, that the engine could be held at the point of valve crash without disaster—and we were told that the engine could be forced *past* the point of valve crash, where little power is being produced, after which it begins to pull again and will continue to gain until a piston breaks or some other mechanical disaster occurs. Most drivers will, we assume, be content with the engine speeds attainable before valve crash starts. The more carefree types will be pleased to know that occasional excursions into the range of valve crash will not result in a valve breakage.

Even though the P-1800 engine's 100 bhp (@ 5500 rpm) and 108 lb/ft of torque (@ 4000 rpm) do not make it one of the world's outstanding power producers, even in its displacement category, it promises to be one of the most reliable. The proof-test for the design was 500 continuous hours of running on an engine dynamometer, pulling full-throttle, full load, all the while. Five hundred hours of pumping out 100 bhp at 5500 rpm. This means that the P-1800 car will be capable of cruising as fast as road conditions, and local speed laws, will allow for almost indefinite periods.

TRANSMISSION

The transmission used in the P-1800 is one that has just recently gone into production and is already in use in the company's sedans. It is *not* just a re-work of the previous 3-4 speed transmission, but an all new design

Sturdy crankshaft runs on five main bearings.

with many improvements over previous models. The most obvious difference between this new transmission and the previous one is the change in ratios. The old transmission had ratios of 3.45, 2.18, 1.31 and 1.0 for the four forward speeds, which made first and second gears real "stump-pullers," with a long jump to third. In the new transmission the ratios are 3.13, 1.99, 1.36 and 1.0, giving a "faster" first and second, and a more normal spacing to third and top.

With a very few parts substitutions, this transmission can be converted into a 3-speed unit, but this conversion is not available except in the sedans. Oddly enough, this 3-speed version is only fractionally cheaper than the 4-speed and the sole reason that it is offered at all is to please the people who don't like to shift gears too often.

The most outstanding design feature of the new Volvo transmission is the extensive use of ball and roller bearings—in fact, the only plain bushing-type bearing used is on the reverse gear idler shaft, which is rotating only when the car is backing up. Even the clutch shaft pilot bearing, which fits into a recess in the rear of the crankshaft, is a ball bearing. These anti-friction type bearings are used not only for the long service life that they provide, but also because they hold very close tolerances and eliminate the looseness that can cause gear noise. While on the subject of gear noise, we might mention that in the P-1800 particular attention has been given to making the clutch housing and the mounting flange portion of the engine crankcase very rigid. Flexing in these areas will cause the clutch shaft to run at a slight angle to the transmission and set up a noise at the cluster-shaft input gear.

All of the forward gears are in constant mesh and are fitted with a very effective synchromesh mechanism. This is of the conventional baulk-ring type, but the mechanism itself in the Volvo design is of such size and strength that the mechanism cannot be beaten, no matter how quickly one moves the lever.

A Laycock de Normanville overdrive unit is available and is coupled on behind the transmission casing, where a plain extension would otherwise go. The overdrive is operational only in 4th gear, as the torque capacity of the unit does not allow it to carry more than just the unmultiplied engine torque.

The rear axle ratio supplied with the P-1800 depends upon whether or not an overdrive is used. Without the overdrive a 4.1 ratio is supplied, giving 18.5 mph per 1000 rpm—with the overdrive, a 4.56 axle is used and with the o.d. ratio of .756, an overall ratio of 3.44 and 22.1 mph per 1000 rpm results. The overdrive option would be most valuable in the western U.S. where the strain-free, high speed cruising given by that combination is the most attractive and most often useful. In the East, the standard gearing will probably suffice.

SUSPENSION

The front suspension is of the conventional short-and-long A-arm type, with rubber-bushed pivots inboard and ball joints at the wheel. There is no anti-dive geometry, but the car is very low and perhaps doesn't require such measures. Neither has there been any effort to raise the roll center, but there is a great thick torsional anti-roll bar linking the front wheels. The roll bar is approximately 7/8 in. in diameter and really does produce a telling effect on the handling of the car. There is a mild understeer at all times—even though the rear roll center is quite high.

The steering gear is as conventional as the rest of the layout. A semi-reversible cam-and-roller steering box, with a ratio of 15.5:1, steers the wheels through a 3-piece track rod. Not exciting, perhaps, but this is one of those places where it is rather hard to improve on orthodoxy.

The rear suspension offers us a perfect example of how a conventional live rear axle should be located. Coil springs are the suspending medium, as is the case at the front, and the axle is located by a series of links which control its motion very closely—something that cannot be said of having the axle fastened to a pair of leaf springs. A pair of pressed steel arms locate the axle in a fore and aft plane and another pair of tubular arms running to struts below the axle casing absorb the twisting reactions to driving and braking torque. Lateral location is provided by a panhard rod: a tubular strut fastened to the axle casing near the wheel hub at one end, and to the opposite side of the car structure at the other end. The rear roll center is determined by the vertical location of this rod.

As this is a touring, not a racing car, the springs and shock dampers are relatively soft. The rate at each front wheel is 85 lb/in., and the rate at the rear wheel is 95 lb/in. We might also mention that aside from minor modifications, all of the suspension components *but* the springs and shock dampers are borrowed intact from the 122-S sedan.

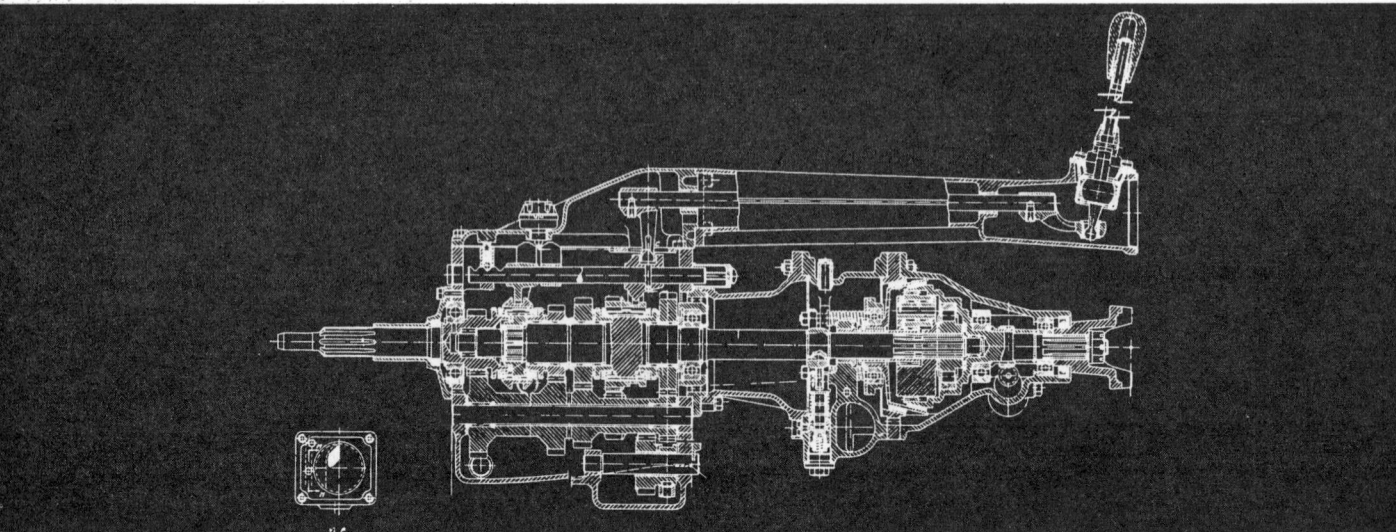

Section view of 4-speed all synchromesh transmission, and overdrive unit.

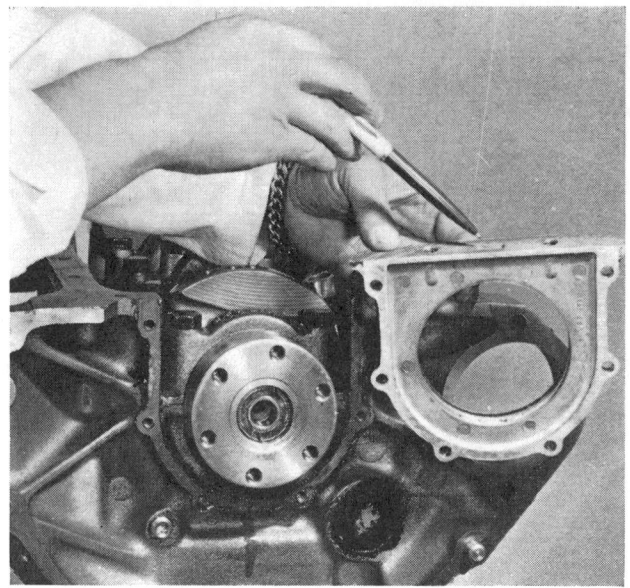
Crankcase is extended well below crankshaft centerline.

BRAKES

The braking system of the P-1800 is a mixture of types using discs at the front and ordinary drum brakes at the rear. The discs themselves are made by Volvo, but the calipers and the rest of the mechanism come from Girling. Volvo's preliminary tests revealed that the exposed in-board pad had a tendency to wear rather fast and a metal shield had to be mounted to keep road dirt from being tossed up on the inboard side of the disc. Drum brakes are used at the rear simply because Volvo engineers felt there was no need for the discs—the rear wheels carry so little of the braking load that discs would have been an unnecessary expense. Moreover, it is just about impossible to arrange an effective hand brake in an all-disc layout and the drum brakes gave an easy solution to that problem.

The lack of the self-servo effect inherent in disc brakes was overcome by the simple expedient of including a power booster in the system. In actual practice, the brakes are smooth, light, and seem to be absolutely free of any trace of unevenness or pulling.

The wheels used on the P-1800 are 15 in. in diameter, just like the ones on the Volvo sedans. However, because low-profile, reinforced tread Pirelli Cinturato tires are used, the rim width is increased to 4.5 in.

BODY/CHASSIS

Like all Volvo automobiles, the P-1800 has a unitized chassis/body structure and, although we have no figures on its torsional strength, it must be very rigid. We drove it over some impossible roads and there was no evidence of twist or shake at all. Volvo was an early experimenter in reinforced glass-fiber body construction, but the P-1800 is all steel. There is an inner shell and undertray, which supply most of the strength, and the outer panels fit over this.

The body contours were developed as a joint effort of Swedish and Italian designers. It is unlikely that any real effort was made to create a "streamlined" package, but the drag readings we took indicated that the car was quite good in this respect.

The actual construction of the body is being done in England under Swedish supervision, and acceptance tests will be carried out at the factory before any car is released to the public. Tests will include the water and air leak tests performed on all Volvo cars.

SUMMARY

The Volvo P-1800 is absolutely orthodox, but has been designed with such attention to detail that it will very likely prove to be an exceptional performer—not in the racing sense, but in that it will provide a lot of trouble-free miles for its owner. Making flat predictions about reliability is, naturally, a very dangerous pastime; however, we do not think that in the case of the P-1800 we are risking anything. The car has a sound basic design and has been given a long and arduous pre-production testing. Further, many of the components were already in use on the 122-S sedan and were well proven at the onset of the P-1800 project. Finally, Volvo's extremely close quality-control methods should effectively eliminate any "lemons." Assuming that you like the looks and "feel," whatever more could you want?

Front and rear suspension details show extensive use of rubber insulation.

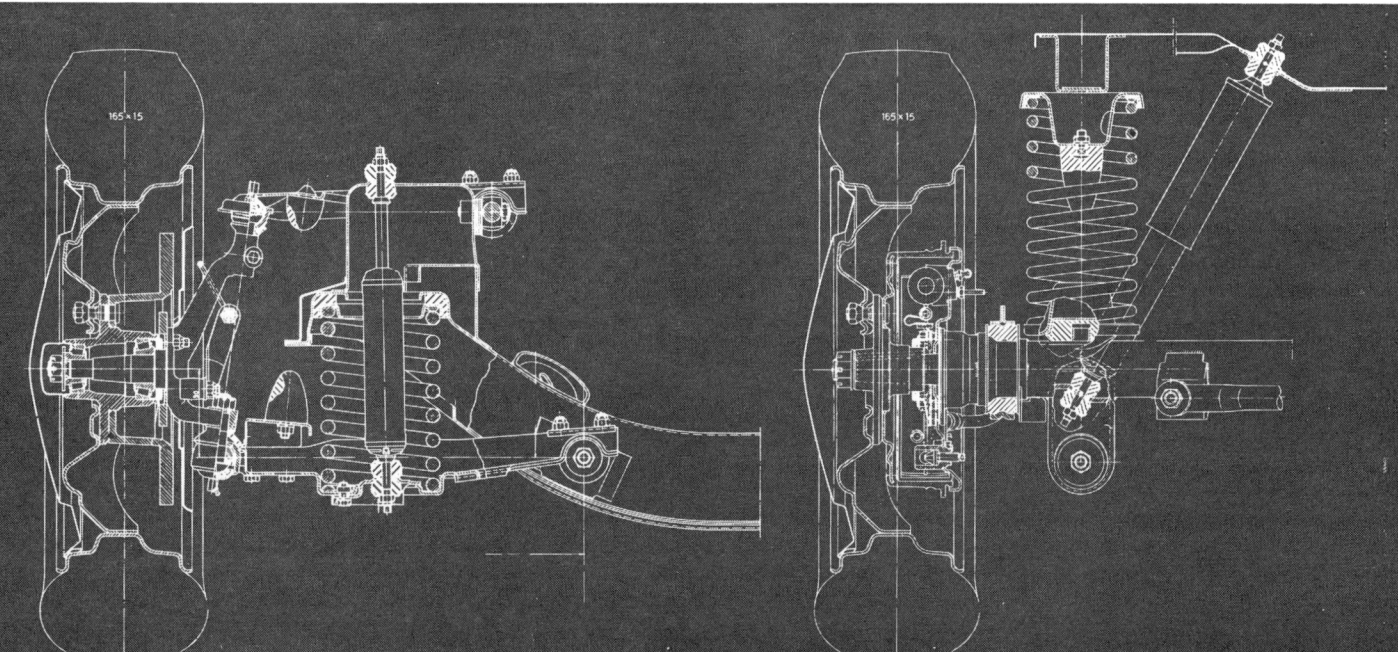

VOLVO 122-S

Quality is its most important asset

OF ALL THE CARS that are brought to our shores, one of the best-suited to American needs and driving habits is the Volvo 122-S. Not because the Volvo is a copy of any American design, but simply because it is sufficiently roomy and has the performance required to cope with our brisk traffic conditions. Moreover, it has the sort of high-speed cruising potential that is absolutely essential for touring on this immense continent of ours.

Of course, there are reasons; all cars reflect the driving conditions that prevail in the countries where they are built, and the Volvo is an excellent mirror for Sweden. Sweden still has a lot of the "wide open spaces" that we fondly imagine to be our exclusive property, and Swedes customarily take extended trips by automobile. At a pretty fierce clip, too, as anyone who has driven there will attest. In fact, they expect much the same sort of service from their cars as we want from ours, and the Volvo is built with that in mind.

In exterior appearance, the 122-S is rather "Ameri-

can" in flavor. It would probably blend right into the general rush and crush of our traffic, were it not for the fact that it lacks the staggering visual impact of some of our stylists' gaudier creations. And, too, in a crowd of the new American cars the Volvo would look a trifle tall. But, though the height of the 122-S may not be modish, it does lend certain advantages to the car. Getting in and out when parked by a high curb can mean a heroic struggle when a contemporary "Detroit" product is involved. But, due to the generous vertical dimensions of the Volvo, one can pop in or out very handily.

Another plus for the car's height is the commanding view it gives in traffic. Naturally, it really isn't all that much higher than any other car, but the extra bit does give an edge in threading through traffic. As a final note on the Volvo's height we would like to say that it is somewhat more apparent than real, the illusion being created by the depth of the unit-structure body. There are no frame members underneath to occupy space, and the car's underpan is also the flooring, which makes the interior seem very deep and the seat very high indeed.

While the styling of the Volvo may have been open to some criticism, the standard of workmanship—in either over-all or detail finish—was not. Our test car was a light pearlescent-grey with a black roof and this gave an over-all effect of subdued elegance. This impression was carried over into the car's interior, where a grey leather-grained plastic covered the most comfortable small-car seats that we have ever seen. The only sour note was provided by the floor mats, also of a grey plastic but very definitely lacking the luxury look of the seats.

The instrument and control layout was very good, with a couple of exceptions. The driving controls, the ones that make the car go, stop and turn, were well arranged and performed their functions precisely and without undue reaching on the part of the driver. And lesser controls, such as the choke, light switch and the like, were also neatly and effectively placed. The instruments, however, although grouped into a neat cluster and very easy to see, gave complete information regarding only the temperature and fuel. Oil pressure and generator charging are covered, *very* loosely, by warning lights. The speedometer was the most uncertain of the lot; it has a horizontal red band to indicate speed and, as this band is slashed off at an angle, one has a fairly broad range of speed, from the point to the heel of the slash, from which to choose a road speed. We used the point, and discovered that it gave very close to the true speed, being just a few tenths slow.

Despite the minor aggravations produced by the instruments (which are much the same in 9 out of 10 cars being sold today) we liked the 122-S. The back seat is, as is the case in all medium to small cars, a bit cramped for 3 persons, but with the armrest pulled down in the middle there is almost armchair comfort for two. The front seats are just about perfect; they are shaped as though the designers had human beings in mind and are adjustable for rake, and for leg room.

Our test car was a bit too new for our liking, being a trifle stiff but, even so, was a real delight. The steering was particularly good; the wheel is placed where it can be cranked around very smartly and the steering gear (cam and roller) gave remarkable precision. Our joy

over the excellence of the steering was, at first, restrained somewhat by the top-heavy feeling of the machine and by the self-steering effects of the rear-axle layout. However, after a time we stopped struggling against these characteristics and learned to use them. Once past this point we had no further difficulties and, in fact, found that the car could be driven very forcefully without the results becoming untidy.

A substantial part of the Volvo's sporting side is supplied by the new 4-speed transmission, which is also used in the new P-1800 sports car, as described in our technical analysis of that car (March 1961). The ratios are much improved over the earlier 4-speed unit—being near perfect—and the synchromesh is absolutely unbeatable. Synchromesh is supplied on all gears and its action is strong enough to allow the driver considerable clumsiness before it makes any audible protest. One thing is certain: what the driver commands, the transmission will have to do, for the shift lever is both long and strong enough to overcome any resistance on the part of the transmission.

Visibility from the driver's seat was good, better than from the older PV-444, but not as good as it might be if it were not for the presence of the unusually thick

window posts. In this respect, however, we can't complain too much; while the thick posts *do* obstruct one's vision slightly, the effect is not so severe as if the top were flattened down level with the hood—which is precisely what may happen when one of those other slender-post lovelies inverts itself. Therefore, any feelings of claustrophobia that we may have had were well mixed with a comfortable sense of being surrounded by protecting structure. This impression of solid strength was materially heightened by the fact that the Volvo never showed the slightest tendency to rattle—even after being subjected to some fast running over a really atrocious road.

Unfortunately, though the body was very solid and little road noise was to be heard, engine noise was very much in evidence. The 122-S is powered with the "sports" version of the PV-series engine and has a pair of SU carburetors (a single Solex is used on the standard model). These are capped with a pair of "gravel-strainer" air cleaners and, while they will prevent large rocks and small animals from being aspirated, they don't hold back the carburetor noise a bit. Therefore, the engine noise varies with the throttle setting, and if the car is being driven very hard, the sound level is entirely too high for comfort. This is an unhappy characteristic and we found it most objectionable. We have been assured that an

effective intake-air silencing system is in the works and will be along shortly.

These criticisms notwithstanding, we were very favorably impressed with the Volvo 122-S. It is a solidly constructed machine, a spirited performer, and handles well. In the area of pure practicality it offers very comfortable seating for four and has a large usable, cube-shaped trunk. Its brakes are good—though not outstanding—and its turning circle allows a U-turn in any but the most narrow streets. Under the hood there is no confused mass of plumbing—everything can be reached and routine service should be easy to perform. And one touch that we liked very much was the radiator-blind; this can be pulled up (by means of a small chain) from inside the car and it gives an extremely rapid warm-up on cold mornings. This last point will be important in far-north areas where the car's occupants will want the heater to start doing its job as soon as possible—and the Volvo's heater/ventilator system will deliver a scorching blast if necessary.

Even though the 122-S has some very tough competitors for the American compact market, Volvo's prospects are quite good. Any shortcomings the 122-S may have are equaled by flaws in its competition, and the Volvo does offer a sporting flavor—perhaps the best quality of all.

ROAD TEST
VOLVO 122-S

SCALE: 10" DIVISIONS

DIMENSIONS

Wheelbase, in	102.4
Tread, f and r	51.7/51.7
Over-all length, in	175
width	63.5
height	59.2
equivalent vol, cu ft	382
Frontal area, sq ft	20.9
Ground clearance, in	7.7
Steering ratio, o/a	n.a.
turns, lock to lock	3.3
turning circle, ft	32
Hip room, front	53
Hip room, rear	52
Pedal to seat back	41
Floor to ground	13

CALCULATED DATA

Lb/hp (test wt)	32.3
Cu ft/ton mile	73.2
Mph/1000 rpm (4th)	16.7
Engine revs/mile	3600
Piston travel, ft/mile	1890
Rpm @ 2500 ft/min	4760
equivalent mph	79.4
R&T wear index	68.0

SPECIFICATIONS

List price	$2495
Curb weight, lb	2400
Test weight	2750
distribution, %	52/48
Tire size	5.90–15
Brake swept area	166
Engine type	4 cyl, ohv
Bore & stroke	3.125 x 3.15
Displacement, cc	1586
cu in	96.6
Compression ratio	8.2
Bhp @ rpm	85 @ 5500
equivalent mph	92.6
Torque, lb-ft	87 @ 3500
equivalent mph	58.4

GEAR RATIOS

4th (1.00)	4.55
3rd (1.36)	6.18
2nd (1.99)	9.05
1st (3.13)	14.2

SPEEDOMETER ERROR

30 mph	actual, 30.9
60 mph	60.7

PERFORMANCE

Top speed (4th), mph	90
best timed run	87.4
3rd (6500)	79.8
2nd (6500)	54.5
1st (6500)	34.8

FUEL CONSUMPTION

Normal range, mpg	24/27

ACCELERATION

0-30 mph, sec	4.8
0-40	7.9
0-50	12.0
0-60	16.6
0-70	23.0
0-80	33.1
0-100	
Standing ¼ mile	20.4
speed at end	66.3

TAPLEY DATA

4th, lb/ton @ mph	160 @ 55
3rd	220 @ 48
2nd	340 @ 36
Total drag at 60 mph, lb	124

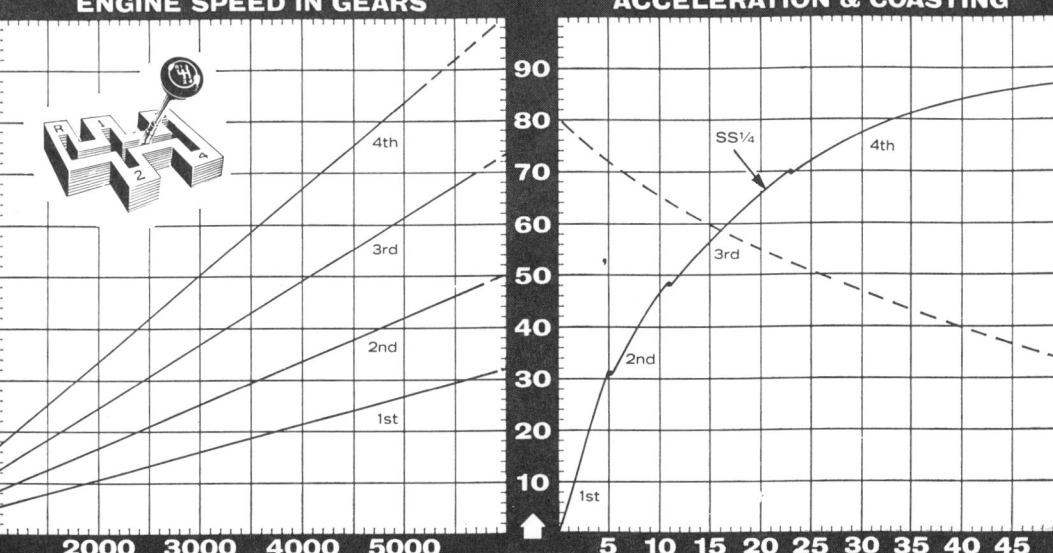

ENGINE SPEED IN GEARS

ENGINE SPEED IN RPM

ACCELERATION & COASTING

ELAPSED TIME IN SECONDS

VOLVO P-1800

![Volvo logo] AFTER MUCH DELAY, the eagerly-awaited Volvo P-1800 Grand Touring coupe is beginning to arrive in numbers on our shores. Full production status for this model was originally planned for early 1961, shortly after we tested the prototype car at Volvo's main plant in Gothenburg, Sweden. But, man's best laid schemes "gang aft a-gley," as the Scot poet once said, and so, too, did Volvo's projected time schedule.

Because Volvo's assembly facilities were even then being strained to capacity in the production of its bread-

and-butter sedans, the job of supplying the unit-constructed body/chassis had to be farmed out. The Jensen Company, in England (which has done a similar job for BMC in the production of Austin-Healeys), was awarded the contract. Jensen subcontracted the panel stamping to Pressed Steel Co. Ltd., and did the assembling in its own plant. All mechanical elements, such as the engine, drive-train, rear brakes (the front brakes are discs from Dunlop) and all but the instruments and some of the miscellaneous electrical components were to be supplied by Volvo's factories in Sweden. Unfortunately, the conclusion of agreements coincided with a spate of industry-wide strikes, with first one group and then another joining in the general fun, and as a result production was delayed for almost a full year.

In this instance the wait was well worth while. The production-form P-1800 retains all of the good features of the prototype—and is better finished and quieter. The styling, which impressed us only mildly in Sweden, is, as it turns out, an absolute smash hit with the American man-in-the-street. Perhaps our tastes have become too astringent to properly appreciate the P-1800's lines because people generally swiveled right around to watch it go by. It was just like those dear departed days when an MG was enough to wow the peasantry, and we would have been something more than human not to have enjoyed the experience.

Closer inspection would not have disenchanted the most sharp-eyed and picayunish of those distant admirers. The color selection for P-1800s is a trifle basic (our test car was an undistinguished off-white) but the quality of the paint and the manner in which it has been applied cannot be criticized. The same thing applies to the body panels (which are fastened together with a proper regard for tolerances) and the trim.

The layout and finish inside the P-1800 are done with a slightly heavy touch but, whatever one might think of the decor, there's no denying that the car is superbly comfortable. The seats—which usually account for most of the comfort, or lack of it, in any car—were of the well padded semi-bucket type. They could have been more deeply contoured, for better support while cornering hard, but were adequate for most touring conditions.

Large doors make it easy to enter the P-1800, and once inside there is a lot more room than is usually found in a car of this type—in every direction but "up"; the roof is a bit low. The seats are adjustable over an extremely long fore-and-aft travel, making it possible for even the often-forgotten tall man to get enough leg room. Also, while doing some fast driving we discovered that there was enough room to permit the driver a great flailing of arms and some determined leaning about without his making a serious encroachment on the passenger's preserves. More leisurely touring allows this space to be used as sprawl-room, and the faint sensation of entombment often experienced in small coupes was missing.

The instrumentation is futuristic, but very complete. The speedometer and tachometer are set into a hooded section right in front of the steering wheel, on either side of a combination mounting that holds the water and oil temperature gauges. There are also gauges for oil pressure and fuel level, and even a clock. Colored lights are present in abundance too, and there is a multitude of small switches for the windshield wipers and washers, heater-fan and so forth.

The only controls not easily seen and reached were the ones for interior lighting, fresh air and the choke. These are tucked away below the dash and gave us some

trouble at first. However, they have coded-shape knobs and after the initial period of familiarization they were not much bother.

Primary controls, such as the steering-wheel, clutch and brake pedals (of the pendant type) and the accelerator pedal, were placed perfectly. So was the gearshift lever, but it had a stiffness—which may not be typical—that countered its nice positioning.

Volvo's safety-harness arrangement deserves special mention. It consists of an adjustable-length strap, anchored to the floor next to the door, that reaches across one's lap to a clip-fastener on the driveshaft tunnel and then travels diagonally over one's body, back to another anchor near the top of the door-jamb. These belts are widely appreciated and used in Sweden and have an honestly-earned reputation there for allowing people to survive most crashes. Apart from this obvious and essential advantage, they are comfortable (and unobtrusive) to wear and easy to fasten and remove. Their only deficiency is in the positioning of the upper anchor, which tends to pull the belt down from the shoulder and onto the arm.

In the trunk, we noticed that there was—by sports car

VOLVO P-1800

standards—an adequate amount of space and that the necessity for carrying a spare tire had compromised what would otherwise have been a very large luggage locker. The compartment is conveniently shaped and could easily accommodate a fortnight's outfitting for 2 if the car's owner could muster enough daring to leave the spare behind in the garage. Lacking this there is, even so, enough room for 1 or 2 medium-to-small bags.

Service accessibility is exceptional. The engine compartment itself is outsized for the amount of machinery it must contain and there is, as a result, very little crowding. Our only adverse comment concerns the location of the battery, which was tucked away up in the right-rear corner where it cannot miss much of the heat from the engine. Should it overflow, there is a strong probability that some of the acid may find its way down onto the passenger's feet.

In the car's various mechanical elements there is considerable evidence of the "strength above all" design-philosophy. The B-18 engine, which will soon be available in the 122-S and PV-544 sedans as well, is a medium-displacement 4, distinguished more by smoothness than vigor and is the most understressed unit we have seen in many years. It has 5 main bearings, and the size of the shaft, bearings and supporting structure seem wasted on only 1780 cc of displacement. Certainly, the engine will take a lot more than is being asked of it at present.

The same thing may be said of the rest of the car. Any place a little bit would have been enough, a lot was used —particularly in the body/chassis structure. In a machine with racing-car pretensions this would be silly, but in the touring-only P-1800 we didn't mind it a bit. Naturally, the Volvo would be faster if it were not carrying all this extra weight around, but the designers have elected, instead, to give us strength and rigidity much above the average—and in that they have succeeded very well.

All cars of this type that come our way for test get a thorough wringing-out on twisty roads and there the Volvo, despite its weight and soft ride, gave a fine performance. There is a dreadful amount of lean while cornering, but the driver can't feel it inside the car and it doesn't seem to affect the handling. Bends, fast or slow, can be taken with *élan*—just a touch of steadying understeer being present at all times.

Acceleration is nothing to get thrilled about. A great many cars less imposing in appearance and less expensive to buy will hand the P-1800 a terrible drubbing in a contest of speed. The clutch bites well, and the engine pulls strongly at all speeds, but the car's weight defeats their best efforts. The Volvo's *top* speed needs no apology, and the gear ratios are spaced nicely for best performance, but the sheer mass of the automobile prohibits sprinting.

In doing that for which it was intended, fast steady cruising, the P-1800 is superb and it gave us the impression it would run forever at near maximum speed. There is little wind noise at high speed and the coil-sprung chassis gives a good ride at any speed and on any surface. No squeaks, rattles, creaks, groaning or road-drumming was heard while we had the Volvo, and we decided that this is a car in which a good radio would not be wasted. Comfort is provided in excellent measure and, with the gearing given by the overdrive (4th gear, direct, gives a "red-line" maximum of 88 mph), the P-1800 is a far faster cruising car than either our road conditions or laws will allow. The price is high, a factor which cannot be ignored, but there isn't another car on the market today that offers precisely the qualities abundantly present in Volvo's P-1800—the potential buyer of a GT-type automobile certainly should see and drive this car.

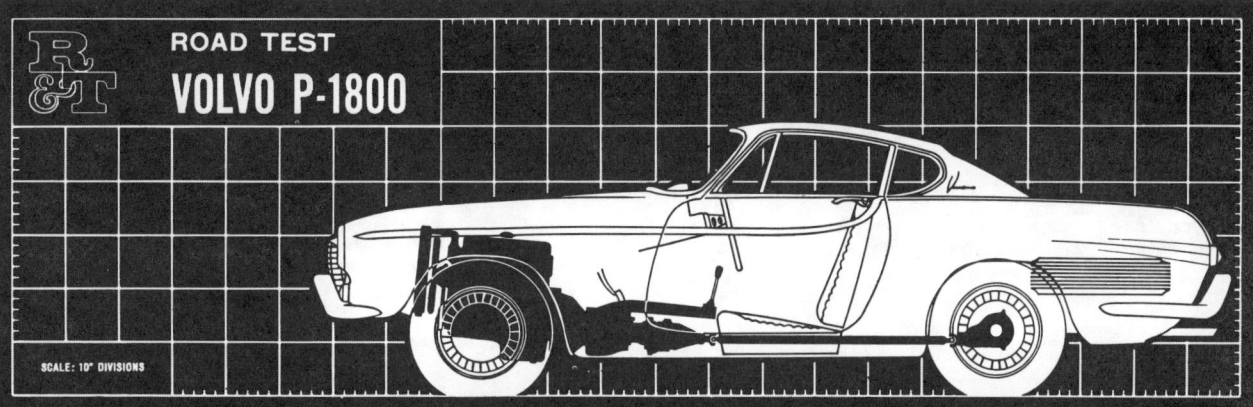

ROAD TEST
VOLVO P-1800

SCALE: 10" DIVISIONS

DIMENSIONS

Wheelbase, in	96.5
Tread, f and r	52.0
Over-all length, in	173
width	67.0
height	51.0
equivalent vol, cu ft	342
Frontal area, sq ft	19.0
Ground clearance, in	5.3
Steering ratio, o/a	15.5
turns, lock to lock	3.5
turning circle, ft	30
Hip room, front	2 x 21
Hip room, rear	50.5
Pedal to seat back, max	43
Floor to ground	10.8

CALCULATED DATA

Lb/hp (test wt)	27.8
Cu ft/ton mile	64.1
Mph/1000 rpm (o/d)	21.2
Engine revs/mile	2840
Piston travel, ft/mile	1490
Rpm @ 2500 ft/min	4760
equivalent mph	101
R&T wear index	42.3

SPECIFICATIONS

List price	$4140
Curb weight, lb	2430
Test weight	2785
distribution, %	54/46
Tire size	165-15
Brake swept area	339
Engine type	4-cyl, ohv
Bore & stroke	3.31 x 3.15
Displacement, cc	1780
cu in	108.5
Compression ratio	9.5
Bhp @ rpm	100 @ 5500
equivalent mph	116.7
Torque, lb-ft	108 @ 4000
equivalent mph	84.8

GEAR RATIOS

O/d (.756)	3.44
4th (1.00)	4.56
3rd (1.36)	6.20
2nd (1.99)	9.09
1st (3.13)	14.3

SPEEDOMETER ERROR

30 mph	actual, 29.8
60 mph	58.4

PERFORMANCE

Top speed (4th), mph	88
best timed run (o/d)	104.7
3rd (5500)	64.5
2nd (5500)	44
1st (5500)	28

FUEL CONSUMPTION

Normal range, mpg	21/28

ACCELERATION

0-30 mph, sec	4.7
0-40	7.1
0-50	10.3
0-60	13.6
0-70	18.7
0-80	25.3
0-100	36.8
Standing ¼ mile	19.0
speed at end	70.5

TAPLEY DATA

4th, lb/ton @ mph	190 @ 55
3rd	260 @ 48
2nd	380 @ 36
Total drag at 60 mph, lb	115

ENGINE SPEED IN GEARS

O.D.
4th
3rd
2nd
1st

2000 3000 4000 5000
ENGINE SPEED IN RPM

ACCELERATION & COASTING

90
80
70
60
50
40
30
20
10

O.D.
SS¼
4th
3rd
2nd
1st

MPH

5 10 15 20 25 30 35 40 45
ELAPSED TIME IN SECONDS

VOLVO 122-S B-18

It may look the same on the outside, but
a new engine with 5-main bearings and disc
brakes on the front wheels are now standard

VOLVO THREE YEARS AGO, when we made our initial acquaintance with the Volvo 122-S 4-door sedan, it impressed us very favorably. Here was a roomy, comfortable, 1.6-liter automobile that was solidly built. And, more important (at least to us), it had very lively performance and a genuine top speed of 92 mph. In fact, that earlier sedan would out-perform most of the new crop of American compacts that were introduced in the fall of 1959.

But the 1959 Volvo had a few minor faults. Specifically,

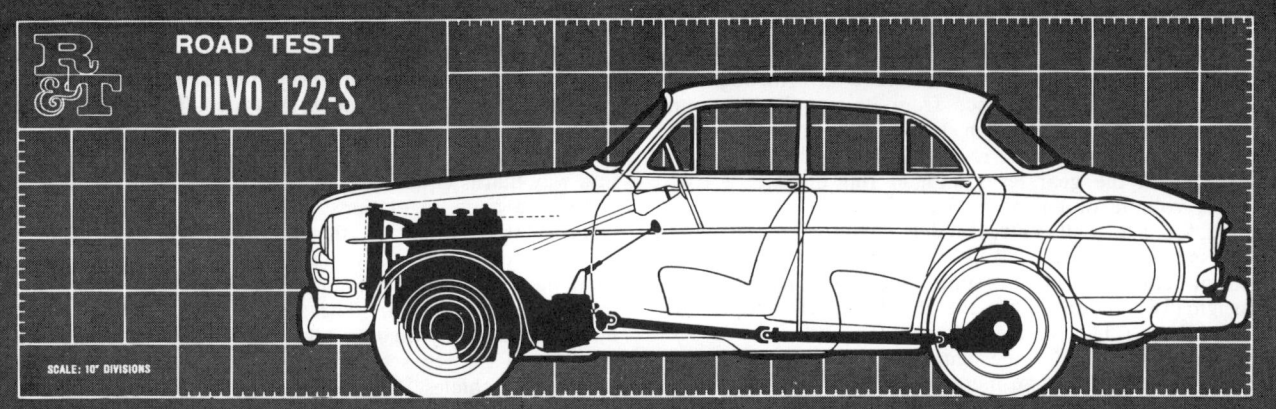

ROAD TEST
VOLVO 122-S

SCALE: 10" DIVISIONS

DIMENSIONS

Wheelbase, in.	102.4
Tread, f and r	51.7
Over-all length, in.	175
width	63.5
height	59.2
equivalent vol, cu ft.	382
Frontal area, sq ft.	20.9
Ground clearance, in.	7.7
Steering ratio, o/a	n.a.
turns, lock to lock	3.3
turning circle, ft.	32
Hip room, front	2 x 19
Hip room, rear	52
Pedal to seat back, max.	44
Floor to ground	13

CALCULATED DATA

Lb/hp (test wt)	30.8
Cu ft/ton mile	74.2
Mph/1000 rpm (4th)	18.4
Engine revs/mile	3260
Piston travel, ft/mile	1715
Rpm @ 2500 ft/min	4760
equivalent mph	87.5
R&T wear index	55.9

SPECIFICATIONS

List price	$2695
Curb weight, lb	2410
Test weight	2765
distribution, %	52/48
Tire size	5.90-15
Brake swept area	339
Engine type	4 cyl, ohv
Bore & stroke	3.31 x 3.15
Displacement, cc	1780
cu in	108.5
Compression ratio	8.50
Bhp @ rpm	90 @ 5000
equivalent mph	92.0
Torque, lb-ft	105 @ 4000
equivalent mph	73.6

GEAR RATIOS

4th	(1.00)	4.10
3rd	(1.36)	5.57
2nd	(1.99)	8.16
1st	(3.13)	12.8

SPEEDOMETER ERROR

30 mph	actual, 29.4
60 mph	59.4

PERFORMANCE

Top speed (4th), mph	93
best timed run	93.5
3rd (5400)	73
2nd (5400)	50
1st (5450)	32

FUEL CONSUMPTION

Normal range, mpg	21/26

ACCELERATION

0-30 mph, sec	3.8
0-40	6.3
0-50	9.5
0-60	14.5
0-70	20.2
0-80	28.3
0-100	
Standing ¼ mile	19.5
speed at end	69

TAPLEY DATA

4th, lb/ton @ mph	205 @ 48
3rd	290 @ 42
2nd	425 @ 37
Total drag at 60 mph, lb	118

ENGINE SPEED IN GEARS

4th
3rd
2nd
1st

2000 3000 4000 5000
ENGINE SPEED IN RPM

ACCELERATION & COASTING

90
80
70
60
50
40
30
20
10
MPH

4th
SS¼
3rd
2nd
1st

5 10 15 20 25 30 35 40 45
ELAPSED TIME IN SECONDS

VOLVO 122-S B-18

the smallish engine had to be thrashed a bit (through the gears) to get full performance from its very-high-output, sports-car-like engine. Though the engine proved that it could really take it, the net result was a certain amount of "rowing" with the gear lever and, at full throttle, a considerable volume of engine noise.

Now, for 1962, all Volvo models including the PV-544 2-door sedan, have a new and larger engine known as the B-18 unit. This powerplant is simply the new 5-main-bearing job originally announced for the P-1800 sports coupe, but de-tuned just a few degrees; down from 100 bhp to 90. With 1780 cc (up 12%), the significant factor is much more torque; now 105 lb-ft—an increase of 20.7%.

This higher torque means much greater flexibility in high gear at low speeds even though the axle ratio has been lowered by 10%. The sum of these changes means less gear-shifting, even better performance and much quieter high speed cruising. Hill climbing ability also improves—our Tapley meter readings show a clear gain of 17% in high gear ability, which simply means that steeper hills can be climbed without shifting down into 3rd gear.

Translating all of these changes into acceleration, nearly a full second has been trimmed from the standing-start ¼ mile time and the car pulls much more strongly in the 40-70 mph speed range usually maintained in touring. An especially attractive feature was the ease with which the car would attain 75 mph in 3rd gear. Needless to say, this makes 3rd a very brisk passing gear: only slightly more than 10 sec are required for the new 122-S to accelerate from 50 to 70 mph. Such performance can certainly be useful in tight situations, even though the B-18 engine still develops a power roar as the engine revs go up. The noise isn't nearly as noticeable as with the old engine, but for those who might object to it, Volvo has an accessory air-intake silencer system, part number 279891.

The 4-speed transmission retains the synchromesh on 1st gear and the ratios are very well chosen. The gears are quiet and, once you've mastered the longish wobble-stick control, it is very easy to select the proper gear for any situation.

The increased performance of the B-18-engined version

of the 122-S has been matched by an improvement in the car's braking power. Dunlop disc brakes, like those on the P-1800, are now used on the front wheels of the sedan. Unlike the sports car, however, the 122-S has no power booster—and apparently none is needed. The pedal pressure required is a trifle higher than before but not enough to be bothersome, and the car will pull down from its top speed in a way that is equaled by few sedans available at any price.

Vehicle code regulations in Sweden are stiffer in some respects than those in the U.S. All cars operating in Sweden, for example, are required to have mud flaps behind each wheel to prevent mud and gravel from being thrown behind the car. The safety conscious Swedes have inaugurated many features we've long advocated for cars and would like to see incorporated, in some form, in all passenger vehicles. The Volvo has a padded instrument panel that is attached to a column built to collapse under pressure, and a plastic package shelf on the passenger's side which folds under impact. The sun visors are of thick foam rubber construction, and safety belts (diagonal straps that extend from the floor in the center across the passenger to the door post) are available on order. The attachment points are installed in every car.

As for the niceties, the early 122-S sedans had somewhat skimpy individual seats in front, but plenty of head and leg room to accommodate the large American physique. Now these 4-door sedans all come through with beautifully upholstered and improved semi-bucket type seats in front and the interiors are definitely more luxurious in appearance as well as in fact. There is room for 5 adults, with a bit of a squeeze for 3 in the rear, but this is a comfortable 4-seater for all practical purposes. In addition to the fine seating, the driver will find that his personal comfort has been considered in the positioning of the controls. The steering wheel is exactly where it should be, there is sufficient room around the pedals to eliminate foot-fumbling, and all knobs and switches are clearly marked and within easy reach. Driving the car is a real pleasure; it is, of course, not up to sports car standards, but relative to other sedans it must be rated as very good.

Seldom does our entire staff reach unanimity of opinion, but in the case of the 122-S it happened. Everyone agreed that this new Volvo really was "superb Swedish engineering." ☉

VOLVO PV-544

If thou shouldst lay up even a little upon a little,
and shouldst do this often,
soon would even this become great.—Hesiod, 720 B.C.

A TIDAL WAVE of cars from across the Atlantic came to our shores about 7 years ago, and in the full flowering of American enthusiasm for the imports, it was difficult to spot those destined to survive. However, few people would have selected the Volvo, a scaled-down 1948 Ford with a smallish 4-cyl engine, as a likely candidate. The Volvo was too dated in appearance, and embodied little in interesting technical features. Far-out engineering was getting most of the play in those days (as is often true now) and the completely straightforward Volvo drew little notice. Nevertheless, after the tides receded in 1960, Volvo was among those which had found a solid and satisfied following of American buyers.

In its original form, the Volvo had the same "1948 Ford bodywork" of the present 544, but was somewhat different mechanically. The early model had a 3-speed transmission and a sports-tuned version of Volvo's old PV-series 3-mainbearing engine, which had bore and stroke dimensions of 2.95 and 3.15 in., and a displacement of 1414 cc.

The new engine, carrying the designation "B-18," also has 4 cyl and a stroke of 3.15 in., just like its immediate ancestor, but that is where the resemblance stops. The B-18 engine has a 5-mainbearing crankshaft, with bearings that are remarkably generous in size; it is strong enough to withstand far more than is being asked of it at present. The block is much roomier than before, and at the present bore size of 3.31 in. there is no crowding. Water completely surrounds each cylinder, and that minimizes thermal distortion. An interesting feature carried over from previous Volvo engines is thermosiphon cooling for the cylinder block. This gives a very rapid warm-up around the cylinders, and that reduces bore wear—which is heaviest when the cylinder-wall temperature is below the dew-point of the corrosive vapors generated in the combustion process.

The cylinder head is blessed with valves and porting that would do justice to a racing engine. All of the ports are separate, and the inlets have inserted rings that perfectly match the manifolding to the ports. The engine is equipped with a pair of SU carburetors. The compression ratio is only 8.5:1, but—oddly enough—at the specified spark setting, the engine would not run on regular-grade fuels without some pinging.

Prior to the change of engines, Volvo had redesigned the old 3-speed transmission into an all-synchro, 4-speed unit: a change that was much welcomed. However, the extra gear was crowded in at some expense in strength, and there were some instances where owner exuberance resulted in the need for repairs. Concurrently with the B-18 engine, Volvo designed and developed an all-new 4-speed transmission with a greater torque capacity and an absolutely unbeatable synchromesh on all forward gears. The gear lever, a long stalk growing up out of the transmission tunnel and inclined back to bring the knob within easy reach, is unchanged. It would be nice (and much appreciated by all of us here) if Volvo would use the transmission extension provided on the P-1800 to bring the lever mounting back nearer the driver, thereby shortening the lever itself, and reducing the "throw" required.

Only detail changes, and exceedingly minor ones at that, have been made in the 544's chassis since its introduction. The front wheels are carried on unequal-length A-arms, and a very light and precise cam-and-roller steering is used. The rear axle, which has hypoid-type gears, is located by trailing links and a transverse track rod. Coil springs and telescopic dampers are used all around.

All of the other Volvos have gone over to disc brakes at the front wheels, but the 544 retains 9-in. drum brakes. Consequently, the 544's braking performance is not as good as the others', but it is still quite good. Our braking tests produced a strong odor of scorched lining, but no perceptible fade.

One of the more attractive features of the 544 is its sturdy and rattle-free unit-constructed body. Window area is a bit limited, as the posts are quite thick, and the styling is neither contemporary nor classic-beautiful, but the use of heavy-gauge sheet steel, and a lot of it, renders the 544 nearly indestructible.

In the interest of making the passengers as bash-resistant as the car, Volvo has developed a seat-belt that is one of the best. It is a strap that starts on the floor, leads across the lap

Volvo: the only way to fly?

to a latch-fitting on the drive tunnel, then goes up and across the chest, and then back to an anchor on the window post.

On the new 544, the instrumentation has been changed to bring it more into line with modern practice, and padding has been added along the top of the dash. The speedometer is now one of those creeping horizontal-line contrivances, and while it may look better than the previous round instrument, it is by no means as readable. The end of the thermometer line is cut on a sharp angle, and one never knows whether to read the point, middle or heel of the slanted end. In checking speedometer error, we used the middle; the error was moderate at that point.

The *circa*-1948 bodywork of the 544 makes for a rather

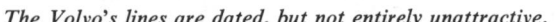

The Volvo's lines are dated, but not entirely unattractive.

ROAD TEST
VOLVO PV-544

SCALE: 10" DIVISIONS

DIMENSIONS

Wheelbase, in	102.5
Tread, f and r	51.0/51.7
Over-all length, in	175.0
width	62.5
height	61.5
equivalent vol, cu ft	390
Frontal area, sq ft	21.4
Ground clearance, in	7.5
Steering ratio, o/a	n.a.
turns, lock to lock	3.2
turning circle, ft	32
Hip room, front	2 x 20.7
Hip room, rear	51.5
Pedal to seat back, max	40.0
Floor to ground	11.7

CALCULATED DATA

Lb/hp (test wt)	27.8
Cu ft/ton mile	82.1
Mph/1000 rpm (4th)	18.4
Engine revs/mile	3270
Piston travel, ft/mile	1720
Rpm @ 2500 ft/min	4760
equivalent mph	87
R&T wear index	56.2

SPECIFICATIONS

List price	$2330
Curb weight, lb	2160
Test weight	2500
distribution, %	52/48
Tire size	5.90-15
Brake swept area	n.a.
Engine type	4-cyl, ohv
Bore & stroke	3.31 x 3.15
Displacement, cc	1780
cu in	108.5
Compression ratio	8.5
Bhp @ rpm	90 @ 5000
equivalent mph	92
Torque, lb-ft	105 @ 3500
equivalent mph	64

GEAR RATIOS

4th (1.00)		4.10
3rd (1.36)		5.57
2nd (1.99)		8.16
1st (3.13)		12.8

SPEEDOMETER ERROR

30 mph	actual, 29.0
60 mph	57.9

PERFORMANCE

Top speed (4th), mph	92
Shifts, rpm-mph	
3rd (5500)	74
2nd (5500)	51
1st (5600)	33

FUEL CONSUMPTION

Normal range, mpg	25-29

ACCELERATION

0-30 mph, sec	4.3
0-40	6.8
0-50	9.6
0-60	14.1
0-70	19.1
0-80	27.0
0-100	
Standing ¼ mile	19.1
speed at end	70

TAPLEY DATA

4th, maximum gradient, %	8.4
3rd	12.6
2nd	19.3
Total drag at 60 mph, lb	150

ENGINE SPEED IN GEARS

4th

3rd

2nd

1st

ENGINE SPEED IN RPM
2000 3000 4000 5000

ACCELERATION & COASTING

90
80
70
60
50
40
30
20
10

MPH

4th

SS1/4

3rd

2nd

1st

ELAPSED TIME IN SECONDS
5 10 15 20 25 30 35 40 45

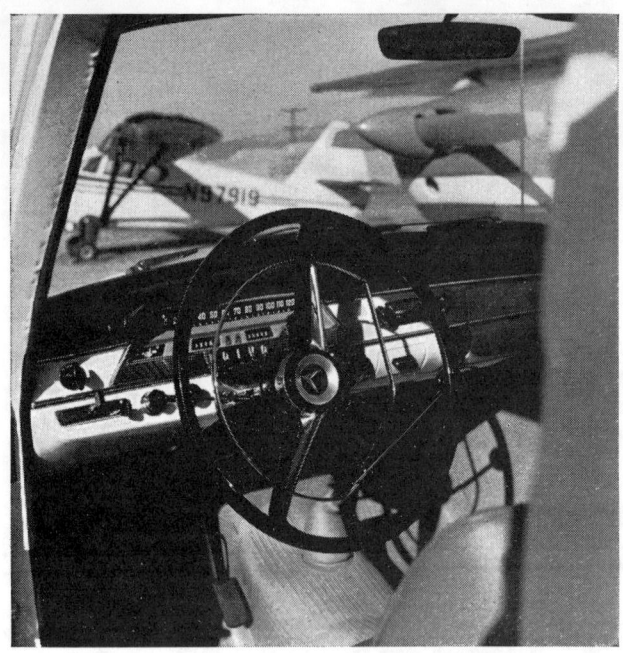

Nicely grouped and clearly labeled controls and instruments.

Accessibility is a requirement that has been fully met.

VOLVO PV-544

narrow interior, but there is adequate shoulder room, and a *lot* of head room. This is one of the few imports that one can drive while wearing a hat—if that matters. Leg room has been supplied unstintingly, but the area around the pedals is a trifle narrow for comfort. The seats are well contoured, and the placement of the controls, relative to the seats, makes this rather a nice car for long trips—much better, in fact, than many another car with nominally more posh interior. The upholstery is all done in a durable and rich-looking polyvinyl plastic, and there are a lot of nice small touches: such as an ash tray at each end of the back seat and back windows that pivot out for ventilation. Everything, except a radio, is included in the basic price of the car—and that includes a venti-

lation and heating system that really does the job as it should.

Trunk room is good by import standards; fair as compared to most U.S.-built compacts: adequate, in any case, for the average family on the average trip (as any married man knows, there can never be *enough* space). At the other end of the car, room has been provided around the engine to make routine service less bother than is so often the case.

Above all, the Volvo 544 is a practical car. Its relatively light weight and small overall size, combined with what is really a very good chassis, make it a pleasure to drive, but its most valuable attributes are economy and durability. True, it cannot match the real midgets for mileage, but it does not have their lackluster performance or limited load capacity, either. If the Volvo has a single most-attractive feature, it is sturdiness and overall quality. There is nothing slap-dash or flimsy anywhere on the car, and this is, in our opinion, more than enough to compensate for any lack of sheer glamour. ⊙

Rather limited visibility astern.

Trunk space is well provided.

VOLVO 122-S AUTOMATIC

The familiar Swedish sedan is now available with automatic transmission

WHEN THE VOLVO 122 was introduced at the London Auto Show in 1957, it was a thoroughly new and interesting sedan that reflected the sound engineering practices and scrupulous attention to quality control for which the Swedish manufacturer was widely respected. These attributes, plus a pleasant appearance and better-than-average road manners, assured the new model a warm welcome. Since that time it has undergone a number of changes that have kept it mechanically up to date and has enjoyed continuing popularity among drivers to whom a car is more than a styling exercise. The latest change is to offer an automatic transmission as an option and this version of the 122-S is the subject of our test.

As it has been four years since we lasted tested a 122-S (the "S" stands for Sport, incidentally, and distinguishes it from the lower-output version sold in the home market), a brief examination of the basic machine is perhaps indicated. The 122-S is offered as a 2-door or 4-door sedan and as a station wagon. In overall size, with a wheelbase of 102.5 and a length of 175 in., the sedan is about the size we think American compacts should be. It is big enough to be practical in U.S. driving conditions, small enough to be easy to drive and yet not so tiny as to be accidentally stepped on. The body/chassis is a welded-up unit and consequently displays both the vices and virtues of this type of construction. On one hand it is strong, rattle-free and durable, but there is also the inevitable kettle-drum effect which results in considerable noise inside even though extra-thick padding is used on the floor.

The front suspension of the 122-S is conventionally inde-

VOLVO 122-S AUTOMATIC

AT A GLANCE...

Price as tested..........................$3015
Engine...............4 cyl. ohv, 1780 cc, 90 bhp
Curb weight, lb..........................2570
Top speed, mph.............................90
Acceleration, 0-60 mph, sec..................15.8
50-70 mph (2nd and 3rd gears), sec..........11.6
Average fuel consumption, mpg...............22

VOLVO 122-S AUTOMATIC

pendent, with A-arms, coil springs, tube shocks and an anti-roll bar. There is a live axle at the rear, but a series of arms and links assures that the axle stays where it is supposed to be and it is consequently far more satisfactory than the average live-axle rear suspension arrangement. It behaves so well, even over rough roads, that it makes you wonder why anyone bothers with independent rear suspension on a front-engine sedan.

Since our last test the engine has been increased in displacement to 1780 cc (from 1586) by enlarging the bore, and there has been an increase in horsepower from 85 at 5500 rpm to 90 at 5000. In design the engine is a completely straightforward 4-cyl ohv with five main bearings and it is carbureted by a pair of 1.75-in. SUs. It is a beefy engine with reserves of ruggedness obviously built in. Other changes in the 122-S include the adoption of the now-popular disc/drum front/rear brake combination and these we found to be fully up to their job.

The driving position is good, the seats are high enough to afford a commanding view of what's going on and are adjustable enough to be comfortable for almost anyone. The steering is quick for a car of this size (3.25 turns lock-to-lock) and its accuracy contributes to the driver's feeling of rapport with the machine.

When the 122-S is driven hard there is considerable body lean and a pronounced understeer, but once the driver has become accustomed to these characteristics it is an easy car to handle at pretty near its limit.

Other features of the Volvo that we like include the over-the-shoulder-and-across-the-lap seat belts that are standard on all models, the impressive care with which everything is put together, and the heater which is one of the most effective in the business. We also heartily approve of the manufacturer's policy of making a genuinely useful range of accessories available. By this we mean that there is not only the usual assortment of sideview mirrors, floor mats, roof racks and convenience baskets, but also that one can obtain such items as a complete service manual ($15), a tourist kit that includes basic spares ($13.19) and even an emergency gas can that fits into the spare wheel ($7.50). Good practical stuff.

The automatic transmission that is now available in the 122-S is the Borg-Warner Type 35, a torque converter with 3-speed planetary gearbox. This is not the finest type of transmission ever built, in our opinion, but it is available to European manufacturers at a reasonable price ($180 more than the manual gearbox in the 122-S) and is adaptable to such widely different machines as the Sunbeam Alpine and the Jaguar 3.8-S sedan. From the enthusiastic driver's point of view, there's simply too big a gap between the three gears, the shifts are relatively slow and, when this transmission is used with a typically small-displacement, low-torque European engine, there is an annoying lurch and a noticeable loss of steerage way after each shift.

We covered a total of about 3000 mi in the 122-S automatic and were able to drive it in conditions that varied from downtown rush-hour creeping to hours of flatland cruising and hundreds of miles over an assortment of mountain roads. Only in heavy downtown traffic could we see any advantage to having the automatic, where it relieved the necessity of rowing through the gears. In highway cruising, where only high gear is used, the automatic was neither a plus nor a minus, but it demonstrated better than average efficiency as we consistently got 23 mpg in this kind of driving. On mountain roads we found the automatic a damned annoyance as it buzzed back and forth from gear to gear and we wished we had a manual box so we could stick it in third and leave it there mile after mile.

We realize that the manufacturer didn't add the automatic transmission to the option list expecting that the experienced enthusiast would become rapturous over it. The automatic is offered because there is an ever-growing segment of the auto driving public that has never learned to use a manual transmission and isn't going to learn. So the manufacturer sells cars that he would not have been able to sell otherwise. It's good business.

But don't let us give you the impression that we didn't like the 122-S automatic. It's just that we think the prospective shiftless buyer is missing part of the fun and pleasure that the 122-S can be.

We must loudly proclaim, however, that a Volvo 122-S with automatic transmission is far better than no Volvo 122-S at all.

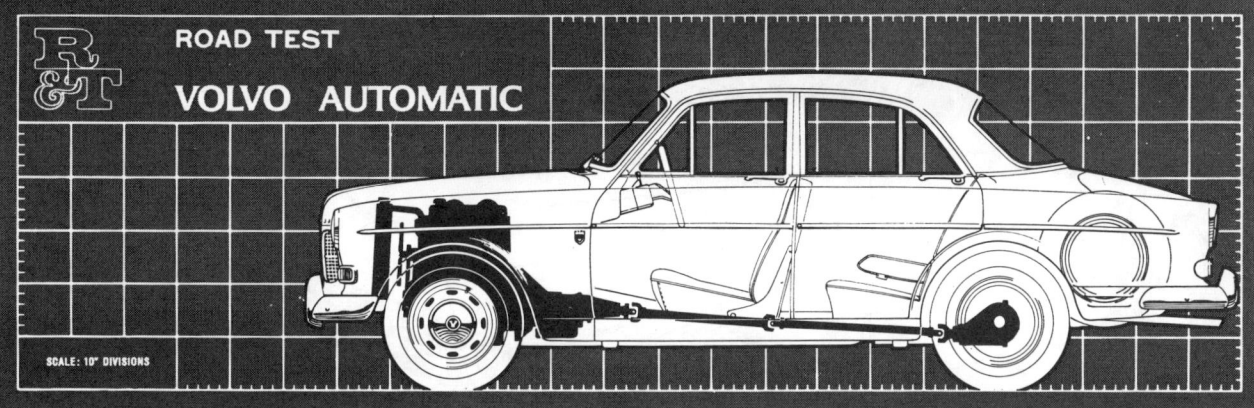

SCALE: 10" DIVISIONS

PRICE

List price $2875
Price as tested $3015

ENGINE

No. cylinders & type . . . . 4 cyl, ohv
Bore x stroke, in. . . . . . 3.31 x 3.15
Displacement, cc 1780
 Equivalent cu in 109
Compression ratio 8.5:1
Bhp @ rpm . . . . 90 @ 5000
 Equivalent mph 88.1
Torque @ rpm lb-ft . . 105 @ 3500
 Equivalent mph 61.7
Carburetors, no. & make 2-SU
 No. barrels & dia. 1-1.75
Type fuel required . . . . . . premium

DRIVE TRAIN

Transmission type: Borg-Warner
 Type 35 automatic (torque con-
 verter with 3-speed planetary
 gearbox).
Gear ratios: 3rd (1.00) 4.10:1
 2nd (1.45) 5.93:1
 1st (2.32) 9.51:1
Converter stall ratio 2.0:1
Differential type . . . . hypoid
 Ratio 4.10:1

CHASSIS & SUSPENSION

Frame type . . . . unit with body
Brake type . . . . disc/drum
 Swept area, sq in 339
Tire size 6.00-15
 Make & model . . . . Goodyear G-8
Steering type . . . . cam & roller
 Turns, lock to lock 3.25
 Turning circle, ft 34
Front suspension: independent,
 coil springs, tube shocks, sta-
 bilizer bar.
Rear suspension: live axle located
 by trailing arms, torque rods,
 and a Panhard rod; coil springs
 and tube shocks.

ACCOMMODATION

Normal capacity, persons 4
Occasional capacity 5
Seat width, front/rear . . 2 x 19/52
Head room, front/rear 42/36
Seat back adjustment, deg. 8
Entrance height, in. 52
Step-over height 13.5
Door width, front/rear 33/29
Driver comfort rating:
 For driver 69-in. tall 94
 For driver 72-in. tall 94
 For driver 75-in. tall 83
 (85–100, good; 70–85, fair; under
 70, poor)

GENERAL

Curb weight, lb 2570
Test weight 2760
Weight distribution (with driver),
 front/rear, % 54/46
Wheelbase, in 102.5
Track, front/rear 51.7
Overall length, in 175.0
 Width 63.75
 Height 59.25
Frontal area, sq ft 20.9
Ground clearance, in 6.9
Overhang, front/rear . . 26.5/44.0
Departure angle (no load), deg . . 14
Usable trunk space, cu ft 9.2
Fuel tank capacity, gal 12

INSTRUMENTATION

Instruments: 120-mph speedometer, water temp., fuel, trip
 odometer.
Warning lights: ammeter, turn sig-
 nal, oil pressure, high beam.

MISCELLANEOUS

Body styles available: 2-door and
 4-door sedans, station wagon.

OPTIONS & ACCESSORIES

Included in list price: 3-point front
 seat belts, heater, vinyl upholstery.
At extra cost: automatic transmission, radio, full range of
 accessories.

CALCULATED DATA

Lb/hp (test wt) 30.7
Mph/1000 rpm (3rd gear) 17.6
Engine revs/mi 3410
Piston travel, ft/mi 1785
Rpm @ 2500 ft/min 4760
 Equivalent mph 84.0
Cu ft/ton mi 77.2
R&T wear index 60.9

MAINTENANCE

Crankcase capacity, qt 4
 Change interval, mi 3000
Oil filter type . . . . full-flow
 Change interval, mi 6000
Chassis lube interval, mi 3000

ROAD TEST RESULTS

ACCELERATION

0–30 mph, sec 5.1
0–40 mph 7.7
0–50 mph 11.1
0–60 mph 15.8
0–70 mph 22.3
0–80 mph 31.7
50–70 mph (2nd & 3rd gears) . 11.6
Standing ¼-mi, sec 20.6
 Speed at end, mph 67

TOP SPEEDS

High gear (5100), mph 90
 2nd (5100) 67
 1st (5000) 43

GRADE CLIMBING
(Tapley data)

High gear, max gradient, % . . . 10.3
 2nd 18.3
 1st 26.9
Total drag at 60 mph, lb 111

SPEEDOMETER ERROR

30 mph indicated . . . . . . actual 27.9
40 mph 37.6
60 mph 57.6
80 mph 79.0

FUEL CONSUMPTION

Normal driving, mpg 20–23
Cruising range, mi 240–275

ACCELERATION & COASTING

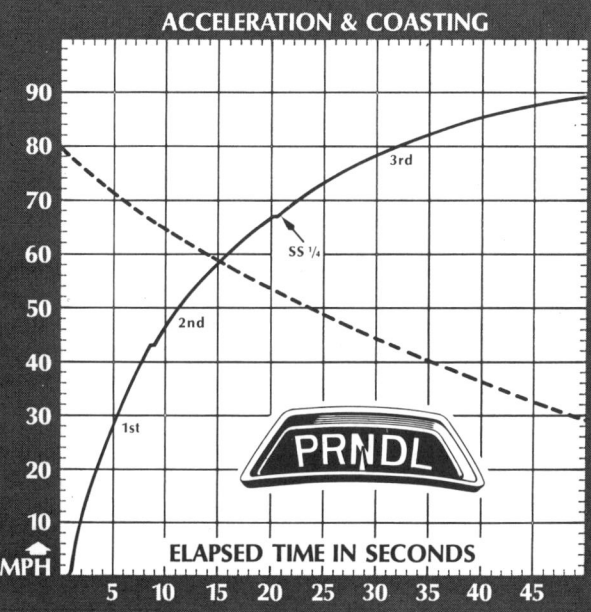

MPH

ELAPSED TIME IN SECONDS

5 10 15 20 25 30 35 40 45

GENE GARFINKLE

AFTER THE NEW WEARS OFF
VOLVO PV-544

BY JAMES T. CROW

W E ARE FORTUNATE to have not one but two cars to examine in this report. The cars are Volvo PV-544s, one belonging to landscape photographer Hans Wendler of Boston, the other to Warren A. Malvick, a computer programmer and engineer who lives in Glen Ellyn, Ill. Both have kept complete records on their cars and the results tell a remarkable story.

The new has been thoroughly worn off both cars. Wendler's PV-544, which was purchased in June 1960, has been used to haul the photographer and his equipment all over the country in all kinds of weather and over all kinds of roads. In the five years he has owned the car he has rolled up a total of more than 133,000 miles and has never had the car quit running or refuse to start. The other car was purchased by Mr. Malvick in May 1959, a year earlier than Wendler, and in six years has traveled more than 78,000 miles. Malvick's Volvo has had widely varied use, one year being driven as many as 22,000 miles and another as little as 3500, depending on the driving the owner's job required.

Maintenance

Both owners have given their cars good care, though it would not be strictly accurate to say that either has been fanatical about observing all factory recommended maintenance to the mile or minute. The Wendler car, for example, has had oil changes and chassis lubes at an average of 2600 miles and tune-ups at an average of 18,000 miles.

There has been an appreciable difference in the cost of maintenance and regular service of the two cars. The Wendler Volvo has required only $364 worth of such services in 133,000 miles while the Malvick car has required a total of $532 in 78,000. This difference can be explained partly by the fact that Mr. Malvick has had the valves replaced (or refaced) at 25,000-mile intervals (as suggested in his owner's booklet) and also needed an out-of-schedule valve job at 62,000 miles. He also has had the greater part of his maintenance performed by an authorized Volvo dealer. About this experience, Mr. Malvick says, "While a member of the Volvo Drivers' Club in Chicago, I found that the best place to take my car for service was to an independent garage. Not only is the work better there but most of the commonly needed parts are always in stock, including rebuilt cylinder heads. Just try to find one of these at any Volvo dealer."

FIG. 2. Malvick Volvo PV-544
Overall Cost per Mile
for 78,000 miles

Delivered price	$2351
Gasoline	1001
Routine maintenance	532
Repairs and replacements	928
Insurance	632
License and taxes	162
Total expenditure in 78,000 mi.	$5606
Retail value at end of period	650
Cost of driving 78,000 mi.	$4956
Overall cost per mile	6.35¢

Fig. 1. Wendler Volvo PV-544
Overall Cost per Mile
for 133,000 miles

Delivered price	$2230
Gasoline	1734
Routine maintenance	364
Repairs and replacements	790
Insurance	1352
License and taxes	325
Total expenditure in 133,000 mi.	$6795
Retail value at end of period	600
Cost of driving 133,000 mi.	$6195
Overall cost per mile	4.66¢

Mr. Wendler's car has been serviced by a Volvo dealer in a small New Hampshire town. "The rate per hour is approximately the same," Mr. Wendler explains, "but much more work is accomplished in the same time. I would not have dared to turn my car over to a Boston dealer at 119,000 miles, leaving it a week, with a 'fix it' carte blanche. My experience with big city dealers has not been too pleasant."

Operating Costs

The operating costs of the two cars demonstrates the considerable difference in costs that can be experienced by two owners of the same model car. Including gasoline, maintenance, repairs and replacements as operating costs, the 133,000-mile car has had a cost of 2.17¢ per mile while the car with 78,000 miles has cost 3.15¢ per mile. We don't have cost figures for other makes of cars with comparable mileage, unfortunately, and therefore hesitate to attempt to make comparisons. We can figure backward on our two Volvos, however (by pro-rating the gasoline and maintenance costs and taking the actual amounts spent for repairs and replacements), and come up with an operating cost for a 36,000-mile period. We do have 36,000-mile costs on other cars to give us a basis for comparison.

Operating Costs[1] of Typical Cars

for 36,000-mile period

Make & model	Cost per Mile
Volvo PV-544 (Wendler)	1.75¢
Peugeot 403	1.84¢
Volkswagen 1200	2.13¢
Ford Falcon 6	2.32¢
Volvo PV-544 (Malvick)	2.42¢
Chevrolet Biscayne 6 w/3-speed manual	2.80¢
Mercury Marauder V-8 w/automatic	4.00¢
Cadillac De Ville w/automatic	4.50¢

From this chart we see that one Volvo was lower in oper-

FIG. 3. Wendler Volvo PV-544

Repairs & Replacements

in 133,000 miles

Timing gear replaced at 65,000 mi.	$ 34
Brakes relined at 72,000 and 108,000 mi.	67
Shock absorbers: 1 set fronts (Konis) at 40,000	42
1 set rears at 60,000	23
1 set fronts at 133,000	22
Exhaust systems, 2 complete, 1 extra muffler	97
Generator brushes at 80,000	3
New fan and belts (3)	17
Two batteries	35
Repair clutch linkage at 72,000 and 130,000 mi.	22
Tires (originals replaced at 43,000 mi, replacements at average of 27,000-mi each)	200
Engine overhaul at 119,000 mi	129
Rubber bushings in front end at 108,000	35
Clean radiator and replace hoses at 102,000	18
Miscellaneous repairs such as flat tires, speedo cable, lights, thermostat, etc.	46
Total repairs and replacements	$790

FIG. 4. Malvick Volvo PV-544

Repairs & Replacements

in 78,000 miles

Voltage regulators at 22,000 and 32,500 mi	$ 33
Speedometer repair at 22,000, 36,000 and 62,000 mi	50
Wheel alignment and balancing in first 22,000 mi.	28
Fuel pumps at 22,000 and 53,500 mi	22
Exhaust systems at 31,000 and 52,000 mi	112
Brakes: relined at 41,000 mi	26
overhauled and relined at 67,500 mi	67
fronts relined at 78,500 mi	20
Carburetor overhaul & tune-up at 42,500	23
Heater repaired at 42,500	36
Batteries at 45,000 and 76,000	41
Tires, 5 new at 56,000 mi	148
Valve job (non-routine) at 62,000 mi	46
Distributor overhaul at 63,500 mi	7
Shock absorbers at 66,000 mi	62
Water pump at 66,000 mi	36
Main bearings and oil pump replaced at 76,000 mi.	79
Throttle shafts replaced at 77,500 mi	20
Front wheel bearing and seals at 78,500 mi	6
Ignition and electrical parts (condenser, lights, starter repair, etc.)	30
Miscellaneous items less than $5 each	36
Total repairs and replacements	$928

ating costs than either a Peugeot 403 or a VW 1200 while the other was slightly more expensive to operate than a Ford Falcon 6 over a 36,000-mile period.

Mr. Malvick admittedly had poor luck during his first 36,000-mile period (see Fig. 3 for repairs and replacement breakdown), his Volvo requiring two voltage regulators, speedometer repairs, fuel pump, exhaust system, wheels realigned and tires balanced during that time. Our other Volvo required very little more than routine maintenance during the first 36,000 miles. The average Volvo owner would probably be less lucky than one but not so unlucky as the other and the typical cost would be somewhere in between.

Overall Cost

The overall cost of the two Volvos is shown in Fig. 1 and Fig. 2. Because of the unusually high mileage we have made a downward adjustment in the retail value of the Wendler car at the end of the period. Wendler's PV-544, with 133,000 miles, had an overall cost per mile of only 4.66¢, which is very low indeed and well below the overall cost of any other car we have examined. It has the greatest number of miles, too, which tends to reduce the overall cost per mile, and it is also five years old, which is past the point of most rapid depreciation.

Figuring backward again, taking an average between the costs of our two examples, and basing the overall cost on an average of 12,000-miles per year for a 3-year period, the overall cost per mile for the Volvo PV-544 would be 7.61¢ per mile. This figure should be more nearly what the average PV-544 owner could anticipate—slightly higher costs than for a VW 1200 or Peugeot 403, appreciably less than for a

Mercury or a Cadillac, and roughly the same as for a 6-cyl, 2-door American sedan like the Chevrolet Biscayne or Ford 500.

Overall Cost[1] per Mile
36,000-mile Period

Make & model	Cost per Mile
Volkswagen 1200	5.13¢
Peugeot 403	6.62¢
Ford Falcon 6	6.67¢
Volvo PV-544	7.61¢
Chevrolet Biscayne	7.96¢
Mercury Marauder V-8	10.00¢
Cadillac De Ville	14.90¢

Depreciation

Though depreciation is not of particular concern to the two Volvo owners whose cars we are considering in this article, the depreciation of the Volvo PV-544 may be of interest to the reader.

Typical Depreciation[2]
Percent of original value lost after

Make & model	1 year	2 years	3 years
Ford Anglia	30%	45%	60%
Hillman Minx	27%	42%	58%
Mercedes Benz 190	19%	32%	48%
Peugeot 403	27%	40%	56%
Volvo PV-544	22%	37%	51%
Volvo 122-S	22%	38%	50%
Volkswagen 1200	05%	16%	32%

These figures show that Volvos, whether the PV-544 or the 122-S, depreciate at a rate that results in about half the retail value being lost in a 3-year period. This is slightly better than the average for all imported sedans, which is about 56%.

Owner Evaluation

The owners of the two cars have been greatly pleased with their cars and both cite dependability and durability as the two greatest assets of the Volvo. Mr. Wendler, who owns the 133,000-mile car, has two suggestions that would improve the car from his point of view. He would recommend that the noise level be reduced, which would make the car less tiring on long trips, and he knows from having tried them that the overall road manners would be improved by using Koni shock absorbers. He has also been annoyed by the car overheating in hot, high country and has replaced the original 2-blade fan with a 4-blade to see if that helps. He also has found that most garages don't yet know how to adjust SU carburetors.

Mr. Malvick also has some suggestions that he feels would improve the PV-544. These include full instrumentation, a larger gas tank, reclining seats, stronger horn ring, radial-ply tires and several other minor items.

PERCENTAGES OF OVERALL COST FOR VOLVO PV-544s

	Wendler PV-544 (133,000 mi)	Malvick PV-544 (78,000 mi)
Depreciation	26%	34%
Gasoline	28%	20%
Maintenance	6%	11%
Repairs & Replacements	13%	18%
Insurance	22%	13%
License & Taxes	5%	4%

Service & Durability

It is interesting to examine the repairs and replacements made on the two PV-544s. Though both have been more nearly trouble-free than would ordinarily be anticipated, both have had things go wrong with them that could not have been prevented by routine maintenance. Of greatest interest, however, is the fact that the two cars shared almost none of the same troubles although they are the same model and virtually identical even though a year apart in age. Wendler's car required a new timing gear after 65,000 miles, for example, while the other is still fitted with the original after 78,000 miles. The oil pump failed in one car after 76,000 miles, which also necessitated replacing the bearings, but the other has encountered no such trouble after 133,000 miles. And so on down the line: No similar troubles at all.

Mr. Malvick has had a continuing problem with the speedometer in his car and has spent a total of $50 trying to keep it working. He finally gave up at about 62,000 and has based his overall mileage on gas consumption since then. Wendler, whose car is fitted with the same speedometer, has done no more than replace a cable. Malvick's car also has needed two new fuel pumps, a heater repair, a water pump and a distributor overhaul which have not been needed on the other car.

A significant item of expense for Mr. Malvick has resulted from the owner's manual recommending valve jobs at 20-25,000-mile intervals. He has had this done at the specified mileages and also needed an additional set of valves at 62,000 miles. On the other hand, Mr. Wendler's *only* valve job came at the time he had the engine overhauled at 119,000 miles.

In general, the 133,000-mile car has given better service than the car with 78,000 miles. But there does not appear to be any correlation between higher mileage and better service. Mr. Wendler says, "I believe my driving is smoother than average, which may account for some of my good fortune. During college, I drove a large, old truck with a non-synchro gearbox in which I hauled a slippery load with no tailgate. I quickly learned how to shift up and down without the slightest jerk."

Undoubtedly, smooth driving does make for greater automobile longevity. It helps to relieve the strain on such components as clutch, gearbox, rear end, etc. But from Mr. Malvick's repairs and replacements (Fig. 4), it seems unlikely that any driver who got 56,000 miles on the original tires (compared with 43,000 for the Wendler car) could be such a rough or uneven driver to have these habits account for the difference in repairs required by the two cars.

It is possible that there is a rational explanation for the difference in overall cost per mile between these two cars—or between any two cars of the same model. Driving habits may have something to do with it, as may the abilities of the mechanic who services the car. We seriously doubt, however, that anything so obvious could make so much difference (except in very gross examples) and will therefore continue to believe, as most drivers do, that there has to be an element of luck involved. If this is true, then certainly Mr. Wendler must be one of the luckiest drivers around.

---

1. Peugeot costs based on report in R&T, June 1965. VW, Ford, Chevrolet, Mercury and Cadillac costs are from "The Real Costs," R&T, January 1964.

2. Based on average retail value of typical 1, 2 and 3-year-old cars in NADA, the National Automobile Dealers Used Car Guide.

VOLVO 1800S

And if 115 bhp isn't enough, there's a hop-up kit that offers more

HAVING BEEN WITH us since early 1961, the Volvo 1800 coupe has attained Old Friend status. A conservative design when it first appeared, it has become comparatively more so over the years, but it still holds a lot of appeal for many people and it's still a soundly constructed, rugged car that offers decent value for the money.

Over its years of production the 1800S has remained very much the same but has been refined in small details, such as a change to small hub caps from the original, over-styled wheel covers; a simpler, more pleasant front bumper; some interior changes including new seats; and two power increases with the latest (as of 1966) bringing the output of the coupe's engine to 115 bhp at 6000 rpm. Another significant change to appear in the 1966 car is a front/rear brake proportioning valve to reduce the chance of rear brake lock-up in hard braking—a laudable improvement that is also applied to the rest of the Volvo line this year.

Getting behind the wheel of the 1800S, one is immediately aware of a rather old-fashioned atmosphere about the seating position. You sit low in the car, the steering wheel is as nearly vertical as any we've tried lately; headroom is a little restricted and the windows seem like narrow slits compared with those in some of the low-waisted cars of today.

This classical feeling in the driving position is visually counterbalanced by an instrument panel that smacks of American Contemporary circa 1955. It's a handsome enough layout but a bit more stylized than many sports car drivers

would care for, with heavy chrome bezels for each instrument and a speedometer that has only 10-mph increments marked out. All the information is there, though—tachometer, speedometer, trip odometer, oil temperature gauge along with the water temp, clock—and in addition to the expected warning lights, a bright red telltale reminds you when overdrive is engaged.

Seats are sufficiently adjustable for fore-and-aft position and rake, and in addition have the novelty of a variable lumbar support. A small hole in the outboard side of the back rest gives access to the large Phillips-head screw that turns a pivoted lateral frame from which are stretched india rubber straps: turning the frame tightens or relaxes the straps to make the seat back stiffer or softer. These seats are covered with real leather and are comfortably soft, vertically and laterally. The lateral softness does not make for sidewise security, but the seats seem comfortable for all, and lateral location of the driver and passenger has been left to the excellent seat belt-body harness arrangement.

Behind the two main seats is a generous luggage area, with leather straps for holding the cargo in place, that doubles as minimal extra seating space. The trunk itself is generous for the size of the car.

Sweden is a country of severe climate, and hence the design of Volvos is influenced by the need for rugged, all-weather cars that will perform reliably and keep the occupants happy in all kinds of weather. Because America is

45

also a land of extremes in weather, the Volvo is better suited to American use than are many imports. There has been more than token attention to rustproofing of the body, for instance, and Volvo heaters are in a class with their American counterparts, as evidenced by a recent test conducted by a Finnish magazine which showed a Volvo 122 to be capable of raising interior temperature from 9°F to over 80°F in 40 min. Too, Volvos have a reputation for living a long time, and their claimed 11-year car life may not be far off.

Sweden also has bad roads in abundance, as do so many European countries. Again the local influence shows to advantage: the 1800S is very much at home on the worst kind of roads with the staunch body structure and fairly soft springing making the road surface a [relatively] negligible factor in determining driving speed. The live rear axle can be made to hop, but the provocation must be severe. On smooth roads where the cornering properties can be evaluated separately, we found moderate understeer, moderate body roll, good directional stability and light, positive steering. Volvo engineers consider radial-ply tires essential to the design and install Pirelli Cinturatos as standard equipment; these combined with tight shock control give the low-speed firmness one expects in this kind of car, but with increasing speed the ride improves. In fact, there are few sports cars with a more satisfactory ride for the long haul. Along with the radial tires, however, comes the slight vibration problem inherent in such tires: our car had a slight resonance coming through the steering wheel at around 70 mph, and careful re-balancing didn't eliminate it completely.

Along with all the rock-solidity comes high weight and modest performance: an 1800-cc engine pulling 2410 lb of curb weight is somewhat restricted by Newtonian physics. The latest power increase shows in the performance figures, but the acceleration of the 1800S is still on the leisurely side. The engine is a most impressive unit, of itself, for a four: vibration periods and power throb are at a minimum and it starts easily, idles smoothly, and runs quietly. To answer

the demand for a little extra, Volvo has worked out a hop-up kit for the B-18 engine and after completing our road test of the standard version we returned it to the distributor's technical center for installation of the kit.

Not a factory-installed option yet, the kit is part No. 419350 and it can be applied to all B-18 engines. For the latest 1800S engine, no changes are required before installing the kit, and it consists of the following pieces:
   -Cylinder head, compression ratio 11.1:1 with 1.65-in. intake valves and 1.38-in. exhaust valves (standard 10:1, 1.97 and 1.65-in. respectively)
   -Camshaft with 0.406-in. lift
   -Lighter flywheel (17.6 lb with ring gear)
   -Oil pump and timing gear covers
   -Pulley hub
   -Carburetor needles, richer than standard
   -Lighter springs for carburetor damper pistons
   -Bosch W280 T 13 S spark plugs
   -Bosch TK 12 A10 ignition coil
   -Sheet metal exhaust header (tuned for extraction)

VOLVO 1800 S
AT A GLANCE...

Price as tested	$4280
Engine	4 cyl in line, ohv, 1780 cc, 115 bhp
Curb weight, lb	2410
Top speed, mph	109
Acceleration, 0-60 mph, sec	13.9
50-70 mph (3rd gear)	8.0
Average fuel consumption, mpg	24.5

ROAD TEST
VOLVO 1800 S

SCALE: 10" DIVISIONS

PRICE

Basic list................$4200
As tested................$4280

ENGINE

No. cyl & type.....4 in line, ohv
Bore x stroke, mm.........84 x 80
 In..............3.31 x 3.15
Displacement, cc/cu in...1780/109
Compression ratio..........10.0:1
Bhp @ rpm.........115 @ 6000
 Equivalent mph............135
Torque @ rpm, lb-ft...112 @ 4000
 Equivalent mph............83
Carburetors..............2 SU
No. barrels, dia.......1 x 1.75
Type fuel required......premium

DRIVE TRAIN

Clutch type......sdp diaphragm
 Diameter, in...........8.5
Gear ratios: od (0.756).....3.45:1
 4th (1.00)........4.56:1
 3rd (1.36)........6.19:1
 2nd (1.99)........9.07:1
 1st (3.13).......14.24:1
Synchromesh..........on all 4
Differential type.........hypoid
 Ratio.............4.56:1
Optional ratios...........none

CHASSIS & SUSPENSION

Frame type.......unit with body
Brake type..........disc/drum
 Swept area, sq in.........339
Tire size.............165 x 15
 Make........Pirelli Cinturato
Steering type.......cam & roller
 Turns, lock-to-lock........3.2
 Turning circle, ft.........31.2
Front suspension: independent
 with unequal A-arms, coil springs,
 tube shocks, anti-roll bar.
Rear suspension: live axle on trail-
 ing arms with Panhard rod, coil
 springs, tube shocks.

ACCOMMODATION

Normal capacity, persons........2
 Occasional capacity.........3
Seat width, front, in.....2 x 18.5
 Rear..............39.5
Head room, front/rear...38.0/28.0
Seat back adjustment, deg......10
Entrance height, in........46.6
Step-over height...........13.7
Door width..............38.0
Driver comfort rating:
 Driver 69 in. tall........90
 Driver 72 in. tall........85
 Driver 75 in. tall........65
 (85–100, good; 70–85, fair;
 under 70, poor)

GENERAL

Curb weight, lb............2410
Test weight..............2770
Weight distribution (with
 driver), front/rear, %....52/48
Wheelbase, in..........96.5
Track, front/rear.....51.8/51.8
Overall length...........173.2
Width................67.0
Height...............50.5
Frontal area, sq ft........18.8
Ground clearance, in........6.0
Overhang, front/rear....31.2/45.5
Departure angle, deg........14
Usable trunk space, cu ft....8.1
Fuel tank capacity, gal......12

INSTRUMENTATION

Instruments: 120-mph speedometer, trip odometer, 7000-rpm
 tachometer, water & oil temperature, oil pressure, fuel level,
 clock.
Warning lights: generator, high
 beam, directional signals, overdrive on.

MISCELLANEOUS

Body styles available: coupe only
Warranty period: 6 mo/unlimited
 mileage

CALCULATED DATA

Lb/hp (test wt)............24.1
Mph/1000 rpm (overdrive)....19.7
Engine revs/mi (60 mph)....3040
Piston travel, ft/mi........1595
Rpm @ 2500 ft/min........4760
 Equivalent mph...........102
Cu ft/ton mi............69.0
R&T wear index..........48.5

EXTRA COST OPTIONS

Radio, limited-slip differential.

MAINTENANCE

Crankcase capacity, qt......4.0
 Change interval, mi......3000
Oil filter type........full flow
 Change interval, mi......6000
Chassis lube interval, mi....3000

BRAKES

Panic stop from 80 mph:
 Deceleration, % G.........68
 Control............good
Parking: hold 30% grade......yes
Overall brake rating........good

ROAD TEST RESULTS

ACCELERATION

Time to speed, sec:	standard	with kit
0–30 mph	4.1	3.8
0–40 mph	6.8	6.3
0–50 mph	10.3	9.4
0–60 mph	13.9	12.7
0–70 mph	18.5	17.4
0–80 mph	24.5	23.2
0–100 mph	51.5	49.0
50–70 mph (3rd gear)	8.0	7.8

Time to distance, sec:
0–100 ft............3.6....3.6
0–500 ft...........10.2....9.8
¼-mile............19.0...18.4
Speed at end, mph....71.0...72.0
Passing exposure time, sec:
 Car ahead going 50
 mph............6.4....6.8

SPEEDS IN GEARS

	normal	with kit
Overdrive mph	109	114
4th (6000)	100	100
3rd (6000)	66	66
2nd (6000)	44	44
1st (6000)	29	29

SPEEDOMETER ERROR

30 mph indicated.....actual 30.0
40 mph.................40.0
60 mph.................58.6
80 mph.................76.9
90 mph.................86.0
Odometer correction factor...1.006

FUEL CONSUMPTION

	standard	with kit
Normal driving, mpg	24.5	23.5
Cruising range, mi	294	281

ACCELERATION & COASTING

ELAPSED TIME IN SEC

KIT
SS ¼
SS ¼ STANDARD
4th
3rd
2nd
1st
¼ MI
1000
500

VOLVO 1800S

In order to install the kit on earlier 1800S cars and B18D engines (122S) certain other preparatory modifications are called for, such as installing correct pistons and/or the oil cooler. The hopped-up engine output is 135 bhp at 6000 rpm, and there is a modest and noticeable increase in performance through the gears without a serious loss in flexibility. Idling speed went up from 700 rpm to 1000 with the kit, but overall noise level didn't increase. The kit will be available through Volvo dealers and will cost $299; installation should run about $100 on current 1800S models.

Staff opinions on the 1800S styling were generally unenthusiastic, with low marks going to the chromium sweep-spear and the semi-finned rear fenders, both cliches of a by-gone American era. The formerly expensive-looking egg-crate grille has been superseded by a stamped affair that doesn't affect the basic styling but does look less elegant.

Good marks go to the shift linkage on this car for precision, easy reaches and low shift efforts. In addition to the good shift linkage, we also liked the method of engaging the overdrive, a handy directional-like lever on the right side of the steering column that's pushed toward the dash to either engage or disengage the overdrive. The unit does not cancel itself when the shift lever is taken out of 4th (the only gear to which the od is applied). Personal taste decrees whether this is an advantage or not, but it didn't seem a bad thing to just leave it on for leisurely driving, going directly from 3rd to 4th od. The overdrive is standard on the 1800S. All indirect gears are quiet and the ratios seem appropriate for the car. Brake, clutch and acclerator efforts are all light and their action smooth, and the whole drive train was tight and slack-free after being attended to by the technical center.

We understand that pops and clunks from the rear suspension and/or axle have been a common occurrence with the model, and our car was delivered to us with a rather severe pop. However, we were pleased to know that the trouble could be (and was) corrected. A small gripe was an oily film deposited on the windshield by the defroster.

To summarize, the 1800S is for the man who wants a solid, conservative and durable car with a little verve and good road manners. Set up properly it gives a feeling of pleasant precision in its responses to driver demands; servicing should be inexpensive, fuel economy is good (24 mpg in all-round driving) and 100,000 miles should be the rule rather than the exception.

NEW FROM SWEDEN

Body of the 144 is its newest component, is contemporary but not radical. Structural rigidity and vision are featured.

VOLVO 144

*Proven components plus new brakes
in an attractive new body*

BY STIG BJORKLUND

THE LONG-RUMORED new Volvo model has been revealed in Sweden and, contrary to all the rumors, it is a 1.8-liter car, not a 2.0 or 2.6-liter. Otherwise, it is pretty much what we expected—a slightly larger sedan with a contemporary, but not radical, body and refinements to the various mechanical systems.

The new model is to be called the 144. It's a 6-window, 4-door sedan on a wheelbase of 102.5 in., the same as that of the 122. Overall length is 7.5 in. longer at 182.5; it's 68.2 in. wide (vs. 63.8) and 56.2 in. high (59.2). The new body is not highly stylized and thus doesn't reflect any fads of the day, but it is neat and modern and should look good for the duration of an expected 10-year production run. It has curved side windows and a fairly low waistline, and the 6-window layout makes for rearward vision appropriate to today's heavy traffic. A 4/5 seater of unit construction, it features great structural rigidity and graduated collapsibility at the ends of the car—a feature of Mercedes-Benz and soon to be a part of domestic Ford design but still the subject of some controversy.

Body

The body sides, including all door jambs and upper and lower longitudinal rails, are pressed as a single piece in the manner of the Rambler "unisides." Windshield-rear window surrounds are also one piece. A new technique of using plastic bonding allows reinforcements to be attached at certain points where welded joints would spoil the exterior surface.

Luggage capacity is claimed to be 14.1 cu ft. by the "SAE Luggage" method; a rather high rear lip contributes to the body's torsional rigidity but will make loading of heavy bags a chore. Front individual seats are fully adjustable for rake and include the lumbar-support adjustment introduced on the 1800S.

Improved door catches are used to keep the doors shut in a crash. The famous Volvo 3-point seat belts are standard on the front seats; there are attachments for them on the outboard ends of the rear bench seat and attachments for a 2-point belt in the center—the first time such an arrangement has been used. The steering wheel hub is recessed and has a large area for load spreading, and the steering column is collapsible by means of a coupling. The bumpers are interesting: shells are of anodized aluminum and have full-width butyl rubber contact surfaces.

In keeping with the demands of Sweden's severe climate, the 144 has a comprehensive heating-ventilation system.

New 144 demonstrates its cornering characteristics at the Volvo proving ground. Rear view, like front, is simplicity itself.

49

NEW FROM SWEDEN

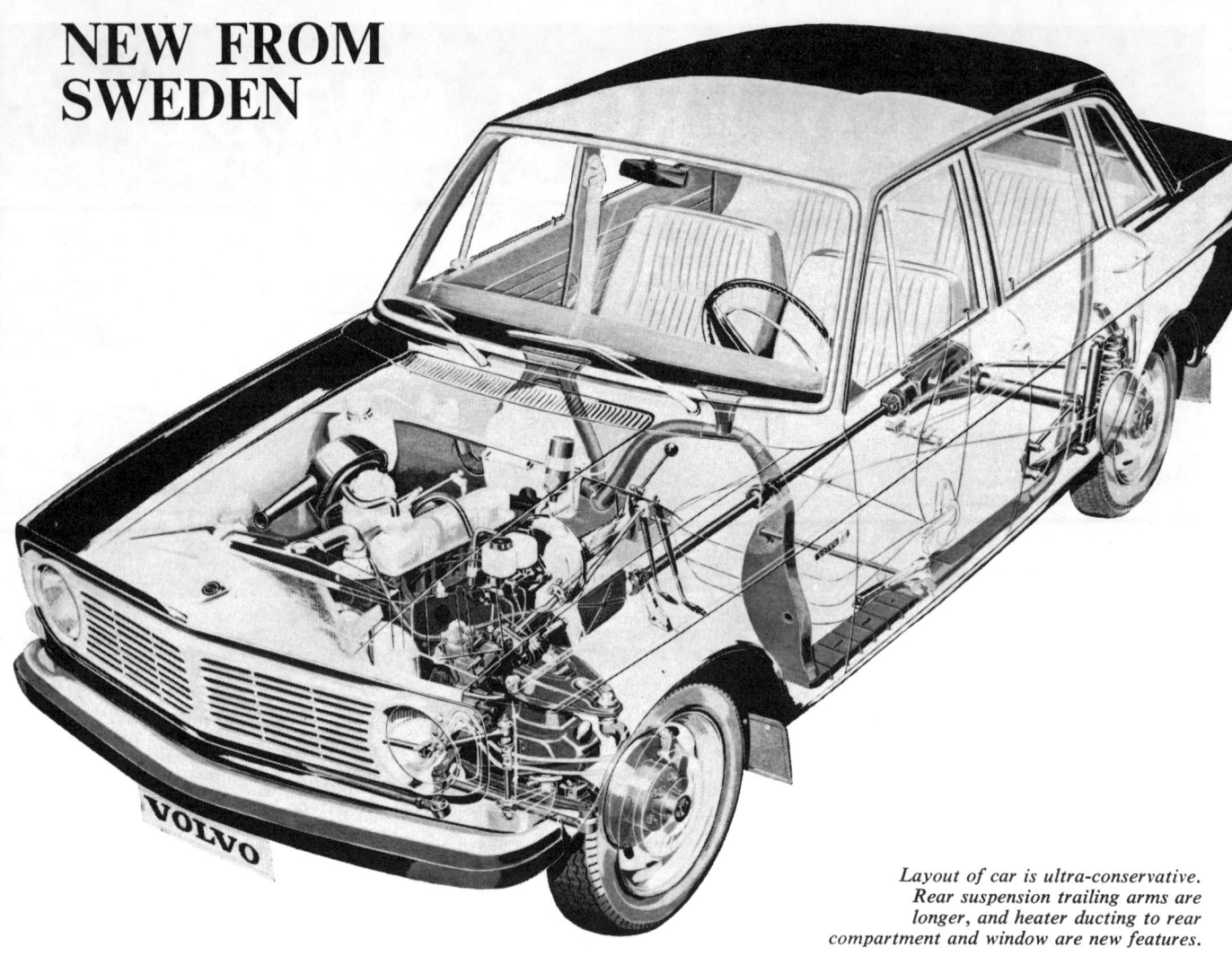

Layout of car is ultra-conservative. Rear suspension trailing arms are longer, and heater ducting to rear compartment and window are new features.

There are two separate air duct systems with separate controls—one to the windshield and rear window, one to the floor. The booster fan operates on fresh air and has two speeds, and a thermostat regulates hot water flow to the system to maintain a constant temperature as set by the temperature control.

Engine & Drive Train

The 121/122S models have had 75- and 95-bhp versions of the B-18 engine, respectively. The 144/144S models have 85- and 115-bhp versions of the same engine, and it's reasonable to assume that only the 144S will be sold in the United States. The standard version is a 1-carburetor unit and the S is the same version used in the current 1800S coupe, with two SU carburetors and a low-restriction exhaust system. New are silencing air cleaners, which should eliminate the objectionable power roar of the 122S. Otherwise, this is the same sturdy, reliable and conservative 1778-cc unit we've known for years.

Clutch, gearbox and differential are unchanged from the smaller car. The Borg-Warner automatic is available on the 144 and overdrive can be had on the 144S.

Suspension & Brakes

Front and rear suspension are conventional and little changed from the 121/122S. There are refinements to rubber isolation, and the trailing arms for the live rear axle are longer. The current 15-in. wheels are continued, with 165-15 tires. Steering takes four turns lock-to-lock and turning circle is 30.3 ft.

The brakes are the most striking mechanical change in the 144. The system is indeed impressive and should set a standard for medium-class cars to equal. There are discs at all four wheels, a double master cylinder and a direct-acting vacuum assist servo. The front discs have double-pad calipers and the rear single. Unusual is the word for the master-cylinder split: each circuit feeds one set of front pads and one rear brake. Thus failure of either circuit leaves 80% of full brake torque and the bulk of the braking at the front, which seems a good idea. Each rear line is equipped with a pressure limiting valve, continuing Volvo practice established earlier this year, and the Volvo people claim there's no chance of rear-wheel lockup with either or both circuits working, and no loss of control with one circuit operative.

The parking brake is a small drum at each rear wheel, not built into the rotor as on the Ate or Chevrolet designs.

In summary, the 144 seems to be a significant new model. It reflects the rising standard of living in Sweden but doesn't forsake the common-sense approach taken traditionally by Volvo. At 150 lb heavier than the 122S it would have been nice to have some extra displacement, but otherwise it seems a logical development of a logical car.

Instrument panel is surrounded by padding and large area of wheel hub is padded. Markers on speedometer are movable.

Joint in steering column breaks under crash impact. Brake booster is above.

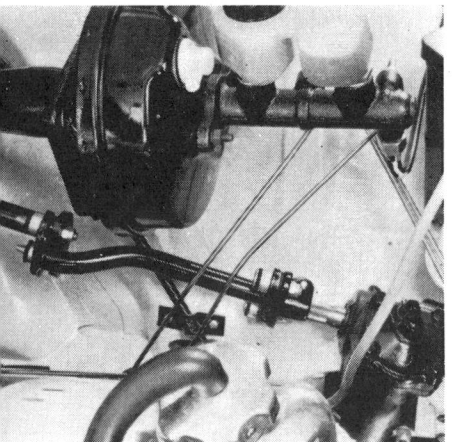

VOLVO 144
Specifications (Sports version in parentheses)

Engine.......................4 cyl, ohv	Tire size.........................165-15
Bore x stroke, mm..............84.1 x 88	Steering.....................cam & roller
Displacement, cc...................1778	Top speed, mph................93 (102)
Compression ratio.........8.7:1 (10.0:1)	Frame....................unit with body
Bhp @ rpm.....85 @ 5000 (115 @ 6000)	Front suspension: unequal-length A-arms,
Torque @ rpm, lb-ft.	coil springs, telescopic shock absorbers,
112 @ 3000 (112 @ 4000)	anti-roll bar
Carburetion...1 Zenith Stromberg (2 SU)	Rear suspension: live axle, upper & lower
Clutch .... diaphragm sdp	trailing arms, Panhard rod, coil springs,
Transmission..4-speed (overdrive optional)	telescopic shock absorbers
Synchromesh....................on all 4	Curb weight, lb....................2550
Final drive ratio......4.1:1 (4.56 with od)	Wheelbase, in.....................102.4
Optional automatic....3-speed plus torque	Track, front/rear...................53.2
converter	Overall length.....................182.5
Brakes....disc; split circuit; limiting valve	Width.............................68.2
to rear brakes; drum parking brake	Height............................56.2

VOLVO 122S, 123 GT AND 1800S

WITH THE INTRODUCTION of the new 144 sedan, Volvo has also made numerous minor modifications to the existing 1800S and 122S and added a high-performance version of the latter, the 123 GT 2-door sedan. The 122S and the 123 GT have a new rear axle linkage similar to that of the 144, with longer trailing arms to keep both wheels on the ground under more severe cornering conditions. Other chassis changes include a twin-branch exhaust manifold, sealed cooling system, modified clutch and an alternator replacing the generator. The 3-point lap and shoulder harness has been redesigned and mounting points added for rear belts. Power of the 122S is now 100 bhp at 5700 rpm.

The 123 GT has the high-output engine of the 1800S, developing 115 bhp at 6000 rpm. Exterior identifying features are iodine fog and driving lights, chrome wheel trim rings and twin fender mirrors. Chassis improvements such as 1800S gearbox ratios, firmer shock absorbers and radial-ply tires as standard equipment put the 123 GT in the rally class, in keeping with Volvo's reputation for rugged roadability. The interior has fully-reclining front seats and a 7000-rpm tachometer atop the instrument panel.

Least changed is the 1800S, with the new sealed cooling system and alternator plus a revised side trim.

123 GT has 115-bhp engine, new grille texture and radial tires.

1800S gets the new grille texture too, plus side trim change.

Volvo rally driver Tom Trana shows the cornering style that is popular on graveled roads in international rallies.

VOLVO 144S

Continues the old virtues
and adds some new ones as well

ONE OF THE reasons Volvo owners have been pleased with their cars is the company's basic conservatism. Volvo makes everything just a little stronger than is absolutely required, uses better materials than are absolutely required, and spends more time testing the components to their limits than is absolutely required. The result is a car that is solid, practical, efficient and long-lived—everything transportation really *ought* to be—but a few steps behind the latest styles and technical developments. So we were interested in seeing what Volvo had done with its first significantly new model since the Amazon (later 122S) of 1956, the just-introduced 144S (the S designates the B 18B twin-carburetor, 115-bhp model; the single-carb, 85-bhp B 18A will not be marketed in the United States).

From the outside, the 144S is an all-new car. The body design is up to the minute and hard to fault. It compares favorably in esthetics and space utilization with the best of the current sedans. Its 6-window superstructure is not unlike that of the Triumph 2000, while the whole car has been likened by a Danish magazine to the very handsome OSI-bodied Alfa Romeo 2600 Berlina. The complete form and all its details have been well thought out. We particularly appreciated the simple, rubber-surfaced bumpers, ice-proof pivoting releases of the door handles and the nicely integrated grille and headlights. The lights are accessible with the hood open and can be aimed in seconds by adjusting several screws, although the left-hand light adjustment is a little close to the forward-mounted battery. The trunk is cavernous—the largest and most usable we've seen on a car anywhere near the size. The car looks tall, especially from the front, though at 56.7 in. it is not any higher than most of its contemporaries. The mud flaps are obviously useful in Sweden and Minnesota but look a bit rural in benign climates.

Inside, the effect of modern, tasteful good sense is maintained. The impressive interior furnishings (all-black vinyl on our test car) give an atmosphere of luxury well above that of past Volvos. The instrument panel is very well de-signed. We don't care much for strip speedometers, but this one is easy to read and reasonably accurate (we found ours about 5% fast reading from the end of the pointer, but just about right reading from the back edge of the diagonal). There is a sliding indicator which can be positioned at any desired warning speed; no lights come on and no alarms sound but it's there to remind one of a limit being exceeded. The speedometer face also includes fuel and temperature gauges, and charging, turn indicator, handbrake, high beam and oil pressure warning lights.

The heater/defroster controls, three rotating wheels in the center of the dash, are an example of functional simplicity. With them, any desired condition can be dialed by fingertip (even with mittens!). The rear window has its own defroster. Just to the right of the heater/defroster controls is a nice flush-fitting ash tray, and beyond that the radio, just too far away to reach with the shoulder harness on. The radio in our test car was an excellent AM/FM unit with superb tone. On the extreme right end of the dash, in front of the speaker, is a passenger grab handle, which says something about the enterprising way Volvos are often driven. Below the dash on the right is a large, lockable, drop-down glove compartment, while in the center, above the transmission, a panel gives access to the fuses. More good thinking.

The front seats are very comfortable, with good lateral support, adjustable backrests (which can go all the way down to meet the rear cushions, making one big bed of the whole interior), and a Volvo exclusive called variable lumbar support. This device consists of a small wheel on the side of each backrest which can be set in "hard" or "weak" position to vary the pressure points—and thus lessen fatigue—on long trips. The much-copied Volvo 3-point seat belt/shoulder harness is retained but we found the quick-release central attachment harder to use (and thus less likely to be used) than the old arrangement.

The rear-view mirror is positioned too high, with the result that only the headliner or else the road immediately behind the car can be seen; it is almost impossible to get a proper view down the road with it. Vision is excellent in every other respect, however; the extensive glass area affords near-360° visibility. The rear seat is not as comfortable as those in front, though a fold-down center armrest adds lateral support when less than three are sitting in back. Because of the relatively high roof, there is plenty of headroom front and rear and entrance and exit are easy. A nice minor touch is the use of sliding coat-hanger hooks; we'd suggest getting two or three extras for each side.

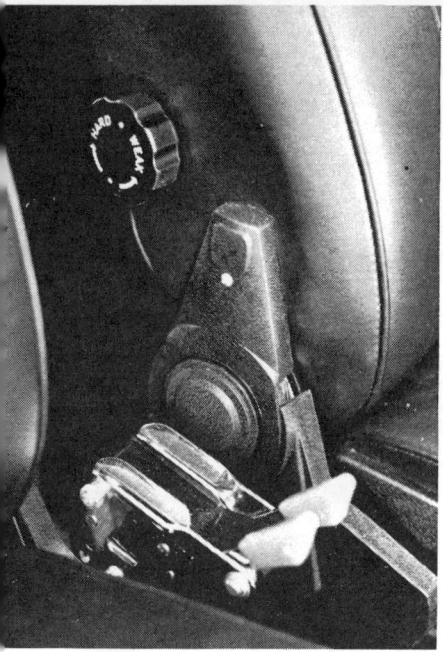

Safety harness now has quick-release catch between lie-down seats. Dial on seat is to select lumbar support firmness.

When one is behind the wheel, the only thing reminiscent of earlier Volvos is the long gearshift lever, the only visible part that is continued unchanged. Once the car moves off, there is no doubt—this is the Volvo of old. The same 1778-cc B 18 engine is still working away strongly and dependably, though it now has the 115 bhp of the 1800S. The engine is both the car's best point and least sophisticated feature. It is responsive, uncommonly reliable and extremely flexible, with good torque over a wide range (though our test car showed a slight initial hesitation on its acceleration runs), and with decent care will go past the 100,000-mi mark before requiring major work (in this connection it is significant to note that the 144S has a 999,999-mile odometer, the only one we can think of—good psychologists as well as engineers, those Swedes!). But the engine is unpleasantly noisy at more than the smallest throttle opening (the familiar Volvo carburetor rasp), though it is reasonably quiet at steady speeds. There is a vibration period just below an actual 70 mph—unfortunately, just the cruising speed many drivers would select. The temptation is to go 75, where it smooths out noticeably, and risk traffic tickets.

The 144S performance figures are much the same as those of the 4-speed 90-bhp 122S we tested in 1962, and the weight is nearly identical, so we are wondering just how much of the extra 25 advertised horsepower finds its way to the flywheel. The B 18 unit first appeared in 1962 as a 5-main-bearing redesign of the old B 14 and B 16 engines, but it would seem to be approaching the end of its development if the refinement of its foremost competitors is to be matched. The B 18 is nevertheless an efficient powerplant, returning about 23 mpg in everyday driving. The 144 has a larger fuel tank (now 15.2 gal) which gives it a cruising range of over 350 miles.

The gearbox is still one of the most satisfying components of the Volvo. Despite the long throws, its action is quiet, smooth and positive, and the ratios are very well suited to the characteristics of the engine. The synchromesh is excellent; the desired gear is always found no matter how carelessly it is selected. Allied to the smooth action of the clutch, the gearbox makes the Volvo one of the easiest cars to drive from the outset. We also tried a 144S with the Borg-Warner 3-speed automatic transmission and found it disappointing. The car seemed powerless at part throttle but roared its throats out with the pedal fully depressed; there seemed to be no way to get a smooth, steady flow of power. In addition, the shift quadrant was not in keeping with the rest of the interior. It looked somewhat cheap and was awkward to use.

The ride and handling of the 144S are hardly distinguishable from its predecessors. Comfort over moderate dips and bumps is good, though on rough roads the limitations of the suspension become evident. On the 144, Volvo has added the longer trailing arms of the 1800S to the familiar live rear axle, coil spring and Panhard rod set-up. It is claimed that these longer arms prevent the inside wheel from lifting even under violent cornering; at any rate, one staff member who

VOLVO 144S
AT A GLANCE

Price as tested	$3225
Engine	4-cyl, ohv, 1778 cc, 115 bhp
Curb weight, lb	2545
Top speed, mph	103
Acceleration, 0–¼ mi, sec	19.6
Average fuel consumption, mpg	23

Summary: Roomier, more attractive version based on familiar, proven running gear ... better brakes ... cavernous luggage space.

VOLVO 144S

has owned two Volvos found the handling and road-holding of the 144S identical for all practical purposes, which is to say very good. Driven hard, the 144S has a basic understeer which can be changed to mild oversteer by getting onto the accelerator. This is a very stable and safe condition for a passenger car since it poses no problems for the average driver yet allows the rally types to get the rear end around. There is no doubt that the car can be cornered very quickly. It will probably be well suited to European-style rallying, as it is every bit as thrashable and uncomplaining under this kind of treatment as were the 544 and 122 series.

Somehow, without increasing the weight of the car or moving the engine forward appreciably (in fact, the longer rear overhang has placed a greater percentage of weight on the rear wheels), Volvo has changed the previously good, light steering into a heavy, insensitive system that almost wants a power assist. The turning circle of the 144 has been reduced by approximately three feet (the wheelbase is unchanged at 102.4 in.), so apparently Volvo has gone for extreme maneuverability at the expense of steering ease. The 144 takes 4.0 turns lock-to-lock compared to 3.25 for the 122S, but the extra three-quarter turn doesn't bring the effort down to an acceptable level. The steering varied so much on the different examples of the 144 we tried that the problem may be compounded by incorrect front-end adjustment.

The brakes are now power-assisted discs all around and

incorporate Volvo's new proportioning system. This has pressure relief valves for the rear wheels and twin circuits arranged so that three brakes still operate should one of the circuits fail. The brakes performed well in our tests, bringing the car to sure, straight-line panic stops and showing no significant increase in pedal effort on successive ½-g stops from 60 mph.

Materials, panel fit, paint and detail finish are all of a very high order. With the minor exception of an unattractive rubber molding at the inner base of the windshield, it is hard to criticize the construction or workmanship of the car. There are no apparent examples of cost-cutting.

All in all, the Volvo 144S is a car of mixed blessings. It has all the solidity and durability for which the company is famous, plus great improvements in styling, interior space and luggage capacity. The larger fuel tank is much appreciated (in this day of credit cards many drivers notice the frequency of gas stops more than they do the actual gallons put in—the Volvo will ease their minds on both counts). The superb front seats contribute to the understated luxury of the interior. It seems to us that Volvo has changed some things that ought to have been left alone (the steering and the safety harness attachments) and left largely unchanged the drawbacks of past models (harsh engine and unsophisticated though effective suspension).

Nevertheless, the Volvo 144S is a far better family car than the 122S, just as that car (which is continued in the lineup) was a great advance over the 544. Enthusiastic drivers may still prefer the 122S, if only for its better steering, but the 144S will please those who want a roomier, more attractive car that is still of moderate dimensions. And of course the big thing with either model is that it will be a long time wearing out.

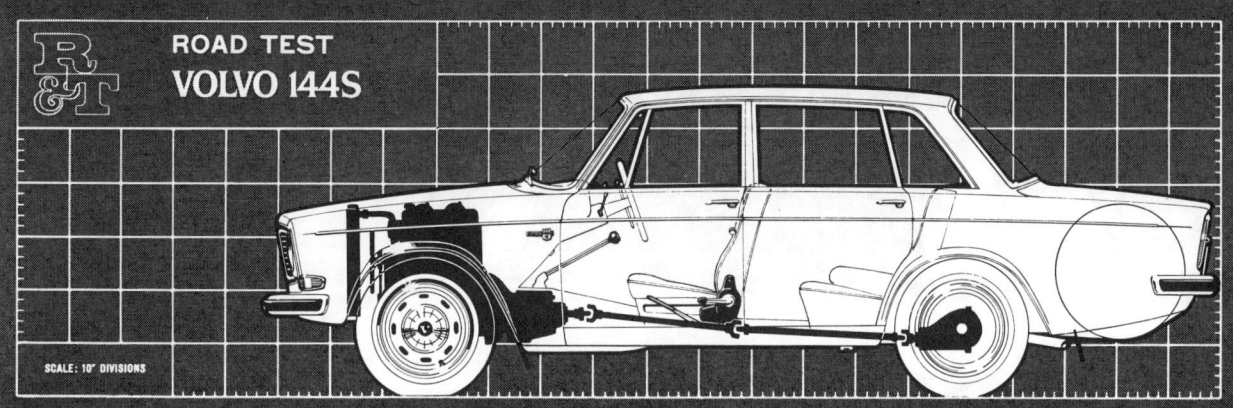

PRICE

Basic list.................$3095
As tested.................$3225

ENGINE

Type.............4 in-line, ohv
Bore x stroke, mm....84.1 x 80.0
   Equivalent in........3.31 x 3.15
Displacement, cc/cu in...1778/109
Compression ratio.........10.0:1
Bhp @ rpm.........115 @ 6000
   Equivalent mph...........109
Torque @ rpm, lb-ft..112 @ 4000
   Equivalent mph............72
Carburetion...........2 SU HS 6
Type fuel required.......premium

DRIVE TRAIN

Clutch diameter, in..........8.5
Gear ratios: 4th (1.00).....4.10:1
   3rd (1.36)............5.58:1
   2nd (1.99)............8.16:1
   1st (3.13)............12.8:1
Synchromesh............on all 4
Final drive ratio.........4.10:1
   Optional ratios (with od)..4.56:1

CHASSIS & BODY

Body/frame.....unit with chassis
Brake type: single caliper solid disc
   front & rear, twin lines, rear-
   brake limit valve.
   Swept area, sq in.........400
Wheel type & size, in....steel,
   15 x 4½ J
Tires..........Firestone 165S-15
Steering type......cam and roller
   Overall ratio............17.5:1
   Turns, lock-to-lock.........4
   Turning circle, ft.........30.3
Front suspension: independent with
   short and long arms, coil springs,
   tube shocks, anti-roll bar.
Rear suspension: live axle with coil
   springs, trailing arms, tube
   shocks, Panhard rod.

EQUIPMENT

Standard: whitewall tires, 3-point
front seat belts.
Included in "as tested" price: AM/
FM radio, outside mirror.
Other: overdrive; Borg-Warner 3-
speed automatic transmission.

ACCOMMODATION

Seating capacity, persons...4 + 1
Seat width, front/rear 21.5 x 2/54.5
Head room, front/rear...40.5/37.0
Seat back adjustment, deg.....90
Driver comfort rating (scale of 100):
   Driver 69 in. tall...........95
   Driver 72 in. tall...........90
   Driver 75 in. tall...........85

INSTRUMENTATION

Instruments: 120-mph speedometer, odometer, trip odometer,
   fuel, water temperature.
Warning lights: generator, oil
   pressure, high beam, directionals,
   handbrake.

MAINTENANCE

Crankcase capacity, qt........4.1
   Change interval, mi.......3000
Filter change interval, mi.....6000
Chassis lube interval, mi.....6000
Tire pressures, psi........20/23

MISCELLANEOUS

Body styles available: 4-door sedan
   (as tested)
Warranty period, mo...........6

GENERAL

Curb weight, lb..............2545
Test weight................2850
Weight distribution (with
   driver), front/rear, %....51/49
Wheelbase, in.............102.4
Track, front/rear.....53.1/53.1
Overall length.............182.7
   Width.................68.1
   Height................56.7
Frontal area, sq ft.........21.4
Ground clearance, in.........7.1
Overhang, front/rear...31.2/49.1
Usable trunk space, cu ft....23.2
Fuel tank capacity, gal......15.2

CALCULATED DATA

Lb/hp (test wt)............24.8
Mph/1000 rpm (4th gear)...18.0
Engine revs/mi (60 mph)....3340
Piston travel, ft/mi.......1755
Rpm @ 2500 ft/min.......4760
   Equivalent mph...........86
Cu ft/ton mi..............73.7
R&T wear index...........58.6
Brake swept area sq in/ton....281

ROAD TEST RESULTS

ACCELERATION

Time to distance, sec:
   0–100 ft...............4.1
   0–250 ft...............6.9
   0–500 ft..............10.5
   0–750 ft..............13.6
   0–1000 ft.............16.4
   0–1320 ft (¼ mi)........19.6
Speed at end of ¼ mi, mph....74
Time to speed, sec:
   0–30 mph...............4.2
   0–40 mph...............6.3
   0–50 mph...............8.8
   0–60 mph..............12.3
   0–70 mph..............17.0
   0–80 mph..............23.5
   0–100 mph.............46.0
Passing exposure time, sec:
   To pass car going 50 mph....8.6

FUEL CONSUMPTION

Normal driving, mpg.......20–25
Cruising range, mi.......305–380

SPEEDS IN GEARS

4th gear (5650 rpm), mph.....103
3rd (6000).................80
2nd (6000).................55
1st (6000).................35

BRAKES

Panic stop from 80 mph:
   Deceleration, % g..........75
   Control..............excellent
Fade test: percent of increase in
   pedal effort required to maintain
   50%-g deceleration rate in six
   stops from 60 mph........14%
Parking brake: hold 30% grade. yes
Overall brake rating.....very good

SPEEDOMETER ERROR

30 mph indicated.....actual 28.2
40 mph....................38.0
60 mph....................57.8
80 mph....................77.3
Odometer, 10.0 mi.....actual 9.96

ACCELERATION & COASTING

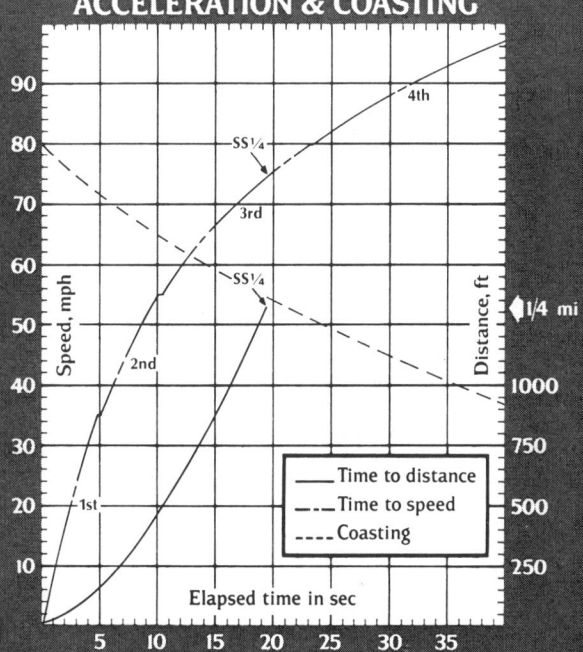

New 164 with 6-cyl, 3-liter engine has same body aft of windshield as 144 but longer front end and new frontal treatment.

VOLVOS FOR 1969

BY STIG BJÖRKLUND

V OLVO HAS GENERATED very little news the last couple years but for 1969 there are a few things to spruce up the existing lines, and the long rumored 6-cylinder car has finally made its appearance.

The 122 models will be sold at reduced prices in the home market to meet expected competition from the upcoming Saab 99, but these dated models have been dropped from the American lineup completely. All cars that had been powered by both the single- and dual-carburetor 1.8-liter engines are getting corresponding versions of an enlarged, 2-liter unit for 1969. This means that the 142, 144 and 145 models sold in the U.S.—the S version—will have the 2-carburetor version.

The new engine is a stretch of the existing 4-cyl unit: stroke remains the same at 80 mm, and the block has been partly redesigned to accommodate a bore of 89.9 mm which raises displacement from 1778 cc to 1986. Bigger intake valves (42 mm vs. 40), freely rotating valves, stronger pistons and rods, and various other minor detail changes have

New 6-cyl engine for 164 uses same pistons, rods, etc., as 2-liter four. Additional cylinders require a longer hood.

Gearshift lever on 164 rises vertically from gearbox tunnel rather than from far ahead as on earlier Volvo sedans.

Volvo 164 Specifications

Price, approx.	$4000	Front suspension: unequal A-arms, coil springs, tube shocks, anti-roll bar	
Engine	ohv inline 6		
Bore x stroke, mm	88.9 x 80.0	Rear suspension: live axle, trailing arms, Panhard rod, coil springs, tube shocks	
Displacement, cc	2980		
Compression ratio	9.2:1	Steering type: recirculating ball, power assisted	
Horsepower @ rpm	145 @ 5500		
Torque @ rpm, lb-ft	163 @ 3300	Curb weight, lb	2830
Carburetion	2 Zenith-Stromberg	Distribution, front/rear, %	54/46
Transmission: 4-speed manual or 3-speed automatic		Wheelbase, in	106.5
		Track, front/rear	53.2/53.2
Final drive ratio	3.73 or 3.31:1	Overall length	185.8
Brakes	discs front & rear	Width	68.4
Tires	165S-14	Height	56.7
Body/frame	steel unit	Fuel tank capacity, gal	15.5

also gone into the larger engine. The extra displacement has not been used so much for extra power as for a fatter torque curve in the low and middle speed ranges: power is up only from 115 bhp at 6000 to 118 at 5800. But peak torque is up from 112 lb-ft at 4000 rpm to 123 lb-ft at the more useful speed of 3500 rpm, and this should do much to correct what we felt was a certain sluggishness in the 144 as compared with older and lighter Volvos.

Alternators are standard on all models and all but the 1800S (which, strangely enough, will not be called 2000S!) will have a viscous-drive fan clutch that limits fan speed to 3000 rpm. Also, all except the 1800S will have a new air intake system for the engine which mixes heated and unheated air before it reaches the air cleaner by means of a thermostatically-controlled flap and duct system, much in the manner of many 1968 American cars. This is not to be confused with the excellent Volvo emission-control intake manifold, which takes fuel and air mixture through tortuous passages past exhaust heat at low speeds for better vaporization of fuel; this system is continued and is now standard on all Volvos, regardless where they are sold. Completing the engine-compartment changes is a higher-pressure radiator cap which lifts the coolant boiling point to around 250°F.

The Borg-Warner Model 35 automatic transmission, optional equipment, has been revised to give a "torque demand" downshift from 3rd to 2nd gear at low road speeds. This means that one can get that downshift at, say, 25–35 mph without kicking the accelerator pedal all the way to the floorboard—a most welcome change.

Inside the 140 series models there is new spun acrylic upholstery which feels like textile but is as washable as plastic and provides ventilation through its pores.

THE BIG NEWS is the 164, but as with most interesting new models from Europe we won't see this one in the U.S.

Heated air-intake system is added to dual intake manifold for further improvement of warmup, fuel economy, emission.

until later—spring of 1969, we understand. The 164 is an extended 144, with a 4-in. longer wheelbase and 3.0 in. more overall length. Its 6-cyl engine uses up not only the extra inches of overall length but also a little of the air space ahead of the fan on the 144, so it's obvious that no attempt was made to maximize the "long hood look." The sheet metal is all new up front and returns to the Volvo frontal appearance of the 1930s—a big square grille with the old Volvo sign diagonally across it.

The engine is both new and not new. Take the newly introduced B20 engine described above, add two more cylinders, and you have it. Of course it's worked over quite a bit, but it does use exactly the same pistons, rods, bearings and so forth. Like the 144S it carries two Zenith-Stromberg carburetors, but its valve timing is a little milder so that its power peak of 145 bhp comes at 5500 rpm and its maximum torque of 163 lb-ft occurs at just 3300 rpm. Still no overhead cam—Volvo engineers think there is even less need for it in this case than in the 4-cyl engine. Like the 2-liter four, the new six uses thermostatically mixed intake air, the emission-control intake manifold and the viscous-drive fan.

A brand new gearbox has been designed for the 164, which has ratios appropriate for the more powerful engine as well as new remote shift linkage which puts the shift lever back where it ought to be instead of so far forward as it is on the 144—thus the clumsy, long lever is gone.

Since the 164 carries some 175 lb more on its front wheels than does the 144, it needed new steering gear. This it got: a new variable ratio, recirculating-ball gear instead of the straight-ratio, worm-and-roller box of the other models. Overall ratio is 18:1 at the straight-ahead position, increasing to about 26:1 out at the locks for parking ease. Volvo's first power steering will be available on the 164 (standard in the U.S.), and has a constant ratio of 15:1; it is a ZF system with careful attention paid to road feel in its design. Brakes are identical with the 144's all-disc system except for larger front pads and continue the unusual fail-safe hydraulic circuit split, which has been extended to all models.

With the standard 6.85-15 tires and 3.73:1 final drive ratio, the 164 is geared for over 19 mph per 1000 rpm; with optional Borg-Warner automatic transmission a 3.31 final drive is used, and this gives over 21 mph/1000 rpm—Americans should feel at home with this gearing, but acceleration might be a bit slow. Overdrive, optional on all Volvo models in their home market including the 164, is unavailable here.

Though its body aft of the windshield is pure 144, the 164's interior is considerably more luxurious, with carpeting, extra sound insulation, leather seat upholstery and, wouldncha know it, phony wood on the dash. There is a fold-down center armrest in the rear seat, which is primarily designed for two people.

In Sweden the basic 164 costs about $800 more than the 144S so that we can estimate its price as around $4000 in the U.S. And if this really is the price Volvo sets for the car, it should sell very well.

2.0 VOLVO 142S

A small change makes a worthwhile difference

BY HENRY N. MANNEY

WHATEVER YOU MAY think out there in Magazine-land, the road tester's lot is not an easy one. Consider the fact that you have to keep in mind what the manufacturer is trying to say, whether he succeeded in the conditions prevailing in most of his markets, whether the car is really suited (willy-nilly) for the *U*S*A*, and whether it is worth spending a lot of the bank's money on if it came down to the crunch.

Likewise, you have to take personal preference into consideration. I wouldn't take as a gift fully 65% of the road test cars that roll into R&T's parking lot (not that anyone has offered), as they are either trashy or else of a subtly outlandish nature. Just like blind dates. The girl is nice, her cuffs are clean, and she doesn't pick her nose or anything, but ugh. Yet another chap will be swept off his feet. Just look around you.

It is like that with the latest Volvo. One of the guys at the

magazine said gee what a dumb car. It's too heavy, the rockers rattle, it leans like a Peugeot 403, it doesn't *do* anything, and in spite of all the clever lifetime advertising, the warranty is only good for six months and so on. I know all that, but happen to like Volvos because you can get in and out like a Christian human being, the seats are comfortable, and in these days of folded-over tin boxes the Volvo is a better finished and sturdier f.o.t.b. Not only that, but as luck would have it I have driven Volvos in some pretty awful weather, including a long spell at minus ∞ Centigrade in Sweden, and everything always works. A lot of cleverer cars don't work when the temperature gets around freezing. What sells Volvos, in effect, is the feeling that like Bed it is a stable place to be.

Volvo's bag has always been to take a solid, if a trifle dull, design and improve upon it gradually. Lots of other makers do this too with more or less success, depending on what the sales and ad boys tell the engineers that the public really wants. The latest 142S, therefore, is nothing but the 2-door 142 with a new 2-liter engine. Following general practice it would have been feasible to bung out the old one a bit further—unless it has gone as fur as it can go—but instead a new oversquare 3.50 x 3.15-in. block was cast up to give the faithful owner a chance to do his four rebores as usual. Engineers tend to leave their signatures on engines and thus the B20 is quite similar to the old one but is a new design with many improvements. For instance, there is a sealed radiator

2.0 VOLVO 142S

AT A GLANCE

Price as tested	$3265
Engine	4 cyl, 1986 cc, 118 bhp
Curb weight, lb	2515
Top speed, mph	103
Acceleration, 0–¼ mi, sec	18.5
Average fuel consumption, mpg	21

Summary: A good car is made better, quieter by a new, bigger engine.

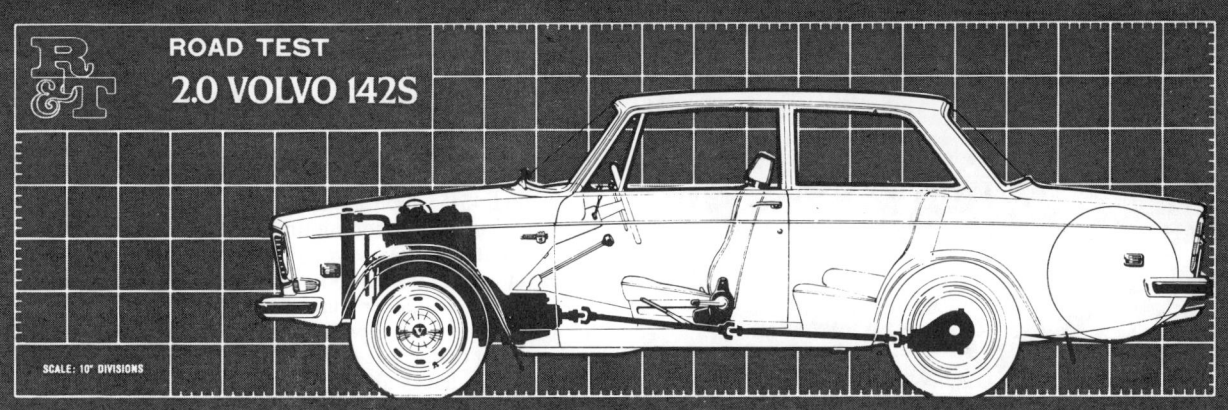

PRICE

Basic list	$3120
As tested	$3265

ENGINE

Type	ohv 4 cyl
Bore x stroke, mm	89 x 80
Equivalent in	3.50 x 3.15
Displacement, cc/cu in	1986/121
Compression ratio	9.5:1
Bhp @ rpm	118 @ 5800
Equivalent mph	101
Torque @ rpm, lb-ft	123 @ 3500
Equivalent mph	60
Carburetion	two 1.75-in. Zenith-Stromberg sidedrafts
Type fuel required	premium

DRIVE TRAIN

Clutch diameter, in	9.0
Gear ratios: 4th (1.00)	4.30:1
3rd (1.36)	5.85:1
2nd (1.99)	8.56:1
1st (3.13)	13.46:1
Synchromesh	on all 4
Final drive ratio	4.3:1

CHASSIS & BODY

Body/frame	unit with chassis
Brake type:	single-caliper solid disc front and rear, twin lines, rear limit valve
Swept area, sq in	400
Wheel	steel disc 15 x 4½J
Tires	Goodyear 165S-15
Steering type	cam & roller
Overall ratio	17.5:1
Turns, lock-to-lock	4.0
Turning circle, ft	30.3
Front suspension:	independent with unequal-length A-arms, coil springs, tube shocks, anti-roll bar
Rear suspension:	live axle with coil springs, trailing arms, tube shocks, panhard rod

OPTIONAL EQUIPMENT

Included in "as tested" price: AM-FM radio ($135), wheel trim rings ($11)
Other: automatic transmission, A/C, limited slip, etc.

ACCOMMODATION

Seating capacity, persons	4+1
Seat width, front/rear	21.5 x 2/54.5
Head room, front/rear	40.5/37.0
Seat back adjustment, deg.	90
Driver comfort rating (scale of 100):	
Driver 69 in. tall	95
Driver 72 in. tall	90
Driver 75 in. tall	85

INSTRUMENTATION

Instruments: 120-mph speedometer, 999,999 odometer, 999.9 trip odometer, fuel level, water temperature

Warning lights: alternator, directionals, brake-on, oil pressure

MAINTENANCE

Engine oil capacity, qt	4.1
Change interval, mi	3000
Filter change interval, mi	6000
Chassis lube interval, mi	6000
Tire pressures, psi	20/23

MISCELLANEOUS

Body styles available: 2-door sedan, 4-door sedan & station wagon
Warranty period | 6 mo.

GENERAL

Curb weight, lb	2515
Test weight	2815
Weight distribution (with driver), front/rear, %	53/47
Wheelbase, in	102.4
Track, front/rear	53.1/53.1
Overall length	182.7
Width	68.1
Height	57.0
Frontal area, sq ft	21.4
Ground clearance, in	7.1
Overhang, front/rear	31.2/49.1
Usable trunk space, cu ft	23.2
Fuel tank capacity, gal	15.2

CALCULATED DATA

Lb/hp (test wt)	24.0
Mph/1000 rpm (4th gear)	17.3
Engine revs/mi (60 mph)	3475
Piston travel, ft/mi	1820
Rpm @ 2500 ft/min	4750
Equivalent mph	82
Cu ft/ton mi	87
R&T wear index	63
Brake swept area sq in/ton	285

ROAD TEST RESULTS

ACCELERATION

Time to distance, sec:

0–100 ft	3.7
0–250 ft	6.5
0–500 ft	9.7
0–750 ft	12.7
0–1000 ft	15.4
0–1320 ft (¼ mi)	18.5
Speed at end of ¼ mi, mph	74

Time to speed, sec:

0–30 mph	3.4
0–40 mph	5.7
0–50 mph	8.5
0–60 mph	12.3
0–70 mph	15.7
0–80 mph	22.2
0–100 mph	44.0

Passing exposure time, sec:
To pass car going 50 mph | 8.2

FUEL CONSUMPTION

Normal driving, mpg	21
Cruising range, mi	320

SPEEDS IN GEARS

4th gear (5900 rpm), mph	103
3rd (6000)	74
2nd (6000)	53
1st (6000)	34

BRAKES

Panic stop from 80 mph:

Deceleration, % g	100
Control	excellent

Fade test: percent of increase in pedal effort required to maintain 50%-g deceleration rate in six stops from 60 mph | 15%
Parking: hold 30% grade | yes
Overall brake rating | excellent

SPEEDOMETER ERROR

30 mph indicated	actual 29.5
40 mph	38.9
60 mph	58.3
80 mph	77.3
100 mph	96.2
Odometer, 10.0 mi	actual 9.75

ACCELERATION & COASTING

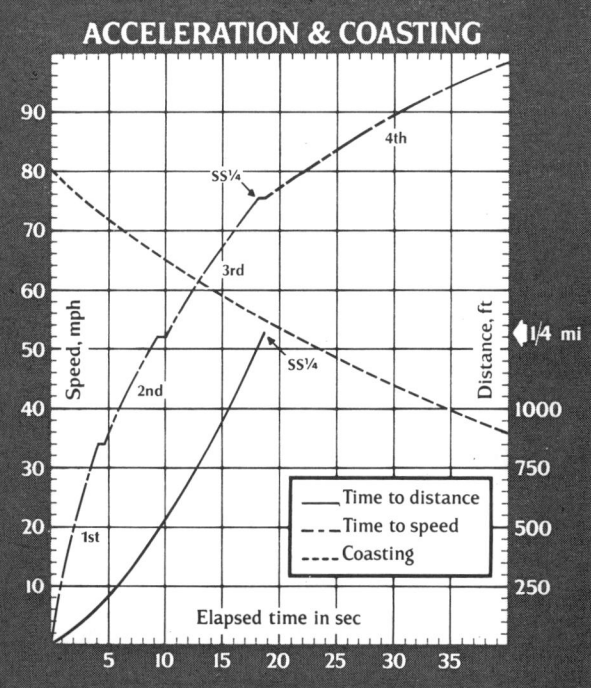

Time to distance
Time to speed
Coasting

Elapsed time in sec

2.0 VOLVO 142S

with increased pressure, a viscous-hub fan to cut down on that annoying row, bigger oil pump, and general beefing up from larger valves to heftier rods. For those of you who remember the B18 engine running on a bit, you will be glad to know that the compression ratio has been dropped to 9.5:1. You will also be glad to know that those zombie SU carburetors have been replaced by similar-looking twin Zenith-Stromberg units of 1.75-in. bore bolted to a trick pre-heated manifold with an eye to the weather in North Dakota, not to mention Green Bay. The induction side is also flattered by a fancy thermostatically controlled cool-air supply with integral flat air cleaner element, thus cutting down on most of the roar characteristic of hungry 4-cyl engines. This increased power is handled by a new beefed-up clutch (which slipped a bit on ours) and flywheel (bigger, by the feel of it) which comes at the end in a higher final-drive ratio (in the stick models) of 4.3:1.

Driving the new model is exactly the same as driving the previous model, except that you are thinking by yiminy why didn't they do this sooner? The extra torque provided by the bigger-bore engine makes its presence felt well down the rpm scale with the result that normal maneuvers such as passing and city-street cornering can be done with a lot more finesse and less "enthusiast" driving. A small by-product of the redesigning is that there seems to be a lot less play between

gearbox and rear axle than has been the case in the past, but the really great advancement is the reduced noise.

The 142S with the B20 is now a really quiet car, thanks to the special fan and air cleaner systems; for really tranquil cruising the rear quarter windows can be opened with a new (but still rattly) catch, the floor air vents opened, and one may motor off with low gain on the radio. The only fly in the ointment really is that there is a definite resonance period around 45-55 mph approximately in top gear, which is a normal non-freeway commuting speed. I don't suppose anyone without rabbit ears would notice it and at any rate, it is better (at least in Calif.) to have it at 50 than 70.

It goes a bit quicker than the 1.8-liter version of the 144 the magazine tested a while back and I refer you to the accompanying data panel for the details on performance.

The Volvo flacks also say that there are some 60 improvements, only three of which are visible. The most apparent is that the famous Volvo seats are now covered with a synthetic cloth instead of leather-like plastic. Although the plastic is pretty good from a wear point of view, I never got used to the smell entirely and besides they are so cold in winter! This B20 arrived with its cloth seats in the middle of our worst hot spell, giving us the feeling of having put on long underwear while summer was still going on, but actually we found that we were a lot less sticky and perspiration-soaked with the cloth seats. Will wonders never cease. It follows that the cloth seats in winter won't give that celebrated imitation of a cold toilet seat like the plastic ones sometimes do either. Another and less fortunate result of the hot weather was that the new carburetors seemed to load up after stopping to park, making it hard to get the engine started again.

We might as well mention that the whole 140 series range of 2-door, 4-door and wagon will be offered with the new engine, automatic transmission with a new kickdown will be available, and even the 1800S coupe will get the 2-liter as well. With increased power, increased torque at the right place, and improved silence of running, the Volvo becomes an even more attractive proposition to those who want to get from place to place in maximum security and comfort.

My neighbor, an airline pilot, took a spin in it and raved about the seats, the phenomenal brakes, the yawning trunk, and the London-taxi-like turning circle. He even said that it went pretty well . . . quite an admission, considering that he drives a 7-liter Brakeless Wonder. "What," he said after a long look, "am I doing driving that and frightening myself when I could be driving this?" Airline pilots are careful people. He may miss the horsepower, but then again he and Volvo may be made for each other. Just like that blind date. ⊙

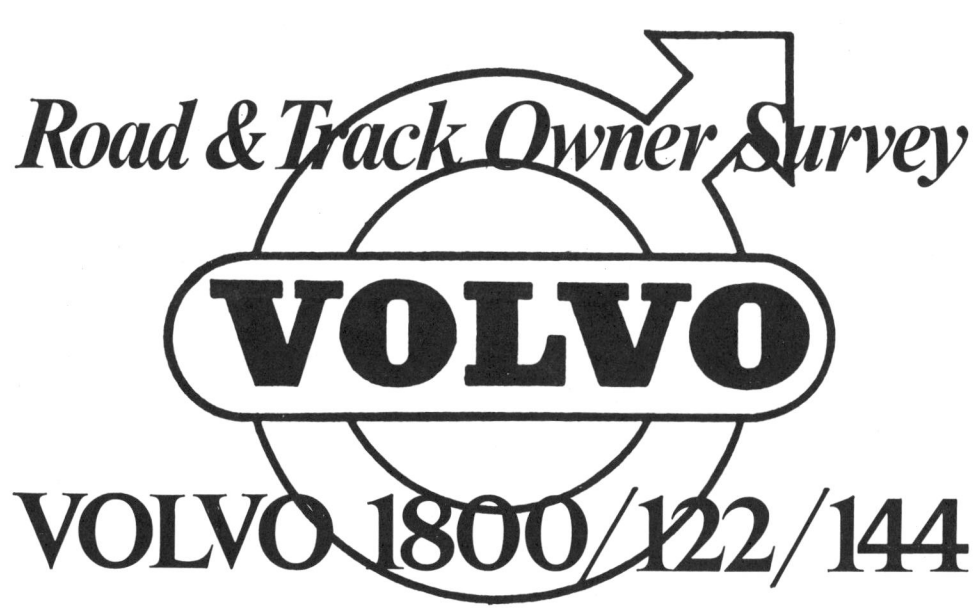

GENE GARFINKLE DRAWINGS

Road & Track Owner Survey

VOLVO

VOLVO 1800/122/144

Our Volvo owners sample didn't include a single 11-year-old Volvo, the average car life claimed in the manufacturer's advertising, but it did contain an unusually high concentration of high-mileage cars and it did indeed show that Volvos are long-lived cars. It also showed that they have at least their share of minor problems to annoy the owner.

There were 134 cars covered in this survey: 77 122Ss, including two of the sporting 123GTs, 29 1800S coupes and 28 of the newer 140 series (the 142S, 144S and 145S wagon). The 122s and 1800s ranged from 1962 to 1968 year models, the

biggest concentrations being 1967s in each case; two-thirds of the 140s were 1968s and the rest 1967s. A recent addition to our questionnaire form is to ask whether the car is a "first" or "only" car or a second (third) car: 80% of the 140s are first or only cars, 68% of the 122s, 64% of the 1800s; the rest are members of a multiple-car family. Eighty-eight percent of all the cars were bought new. Another recent addition to the form covers the owner's age, occupation and the state where he lives; the greatest number of our owners, by a wide margin, fell in the 21–30 age bracket and the three most frequently mentioned occupations were engineer, student and technician.

61

New York was the most represented state, California second; Massachusetts and Florida tied for third.

We eliminated those cars with less than 5000 miles on their odometers from the report. Fourteen percent of the balance had between 5000 and 10,000 miles, 26% between 10,000 and 20,000, 19% between 20,000 and 30,000, 16% between 30,000 and 40,000, and 14% between 5000 and 10,000. Fully 5% had between 80 and 90,000 miles on them and the highest mileage reported was 104,000. Volvo owners use their cars at about the national-average rate—40% of them travel between 10 and 15,000 miles per year and 36% drive between 15 and 25,000 mi/yr. Ninety-five percent of the cars in the survey are used daily for transportation, 68% of them driven on long trips. Sixteen percent of the owners participate in rallies—21% of the 1800 owners, 16% of the 122, and 14% of the 140 owners; 9% of the total drive in a slalom occasionally. The largest percentage in our survey series to date—56%—say they drive their cars "hard," while 37% drive "moderately" and 7% "very hard."

The fact that Volvo owners tend to drive vigorously is reflected in the reasons they give for having bought Volvos. The most frequently noted reason was durability—39% of them mentioned this legendary Volvo quality. Couple that with the 26% who gave reliability as a decision factor and the 7% who mentioned the "Volvo reputation" and you have a strong indication that Volvo advertising and word-of-mouth are convincing people that Volvos are indeed sturdy, reliable machines. This also means that once the owner has bought his Volvo he expects it to be relatively trouble-free. To quote one, "After having supported a Morgan Plus 4, an MGA Twin-Cam and a Porsche 356, I was duly impressed with Volvo's tank-like qualities."

Handling was next on the list of reasons for purchase, and this factor is getting to be a fixture on these lists already; it seems that nearly anyone buying an import expects it to handle well. Naturally, a greater percentage of the 1800 owners considered handling that important: 34%, vs 32% for the 140 buyers and 21% for the 122s. Volvo also has a reputation for good assembly and finish, and 22% of the owners considered this fact an important reason to buy one. Twenty-one percent also were looking for good fuel economy, 20% for safety (another part of the "reputation") and 19% for a high level of comfort. Performance allied to economy, handling and comfort was considered a Volvo combination by 20% of the 122 owners, 14% of the 1800 owners and 11% of the 140 owners before their purchase. The highest previous-ownership percentage to occur so far in this series goes to Volvo: 13% of the owners reporting had already owned one or more Volvos. "Engines and drivetrains absolutely reliable—all could be driven vigorously with little fear of breaking anything"— owner of 145S and two previous 122Ss. Miscellaneous reasons for buying Volvo: styling (mostly 1800 owners), value for money, family space (122 and 140), engineering.

The care Volvo owners give their cars is, as far as we can tell, about average for an imported car owner; 67% follow the maker's maintenance schedule closely, 25% follow it "mostly" and 8% wait for something to break. Once again a "highest yet" percentage goes to Volvo: 14% of these owners do most or all of their own maintenance, and that's a lot. Perhaps this has something to do with their opinion of the dealers; only 50% of the owners consider their dealer's service to be good. (In previous surveys we have found that 62% of VW owners, 56% of Porsche owners and 39% of MGB owners gave their dealers a "good" rating.) Twenty-two percent of the Volvo owners said their dealer service was "fair" and 19% rated it as "poor." As with other makes, about 10% felt that service was too expensive.

New or Used
Bought new . . . 88%
Bought used . . 12%

Miles per Year
0-5000 miles 1%
5000-10,000 . . . . 17%
10,000-15,000 . . . 40%
15,000-25,000 . . . 36%
Over 25,000 . . . . . 6%

How Owners Feel about Volvo Dealers' Service
Rated "Good" . 50%
Rated "Fair" . . . 22%
Rated "Poor" . . . 19%
No local dealer . 6%
No opinion 3%

Do own work . . 14%

About Driving Habits
Drivers who said they drove "Moderately" 37%
Drivers who said they drove "Hard" 56%
Drivers who said they drove "Very Hard" 7%

Factory Maintenance Schedule Followed?
Owners who followed schedule completely 67%
Owners who followed it mostly but not totally . . . 25%
Owners who didn't follow it at all 8%

Problem Areas
Mentioned by more than 10% of the owners:
  Instruments (mostly 1800s)
  Cooling system
  Body parts
  Clutch
Mentioned by between 5 and 10% of the owners:
  Differential   Carburetors
  Oil leaks   Exhaust system
  Wheel alignment Running-on (mostly 1968)
Owners Reporting No Troubles 10%

How Many Current Volvo Owners Would Buy Another?
Would 83%
Would not 12%
Undecided 5%

Five Best Features
Handling
Comfort
Fuel Economy
Reliability
Durability

Five Worst Features
Lack of Power
Ventilation
Engine Noise
Idiot Lights
Harsh Ride

Road & Track Owner Survey
VOLVO 1800/122/144

Of the qualities the owners liked best about their Volvos, 42% listed handling as a best feature—53% of the 140 owners, 48% of the 1800S and 35% of the 122S owners. Comfort came next on the best-liked list: 41% mentioned it, with 19% specifically giving the credit to Volvo's seats. Surprisingly, those with Volvo's unique lumbar-support adjustment in the front seats (1800 and 140 series) didn't brag any more about the seats than those without.

Fuel economy was the next favorite feature, mentioned by 24%; reliability garnered a 23% mention and so did durability. Fewer owners were impressed by the quality level (17%). Other favorites mentioned were performance, solidity, brakes, high-speed cruising ability and overall safety design.

And what's worst about Volvos? Well, for one thing, we have learned in these surveys that owners are better at picking out "worst" features than they are at naming "best" features. Every owner has some little annoyance, and there were 79, count 'em, different "Worst Features" listed by owners, few of them in any significant quantity. All this proves is that nobody can build a car to please everybody. The grievances that do count are: lack of acceleration, 13% (the new 2-liter engine should improve things here); poor ventilation, 10% (the 140s have better provisions); engine noise, 9%; 8% of the 122-140 owners disliked the "idiot light" dashboard and 9% of the 122 owners thought their cars rode harshly. Seven percent of the 122 owners disliked the rear suspension because it clunks, and 17% of the 1800 owners thought the Smiths instruments in their coupes were a worst feature.

As with MGBs and Porsches, Volvo owners reported that instruments were the most frequent problem area, with 17% of them having trouble. Predictably, the Smiths instruments in the 1800S were the biggest offenders—41% of these owners had trouble of some kind with them. Only 10% of the 122-140 owners had instrument trouble, mostly with speedometers or speedometer cables. Next on the trouble list came the cooling system, 13% of the owners having had some trouble here—8% had water pump trouble specifically. Body parts accounted for trouble for 12% of the owners, and the 140 series, relatively new to production, had a 21% incidence of this sort of thing. Window-winding mechanism was the most frequent body trouble in the 122s; vent windows in the 140 owners were troublesome, again indicating some early production difficul-

ties. Differential trouble and oil leaks (mostly from the gearbox) gave 10% of the owners difficulty, and clutch trouble reported by 12% of the owners was in part due to the gearbox leaks. Sixteen percent of 122 owners had some complaints about their exhaust systems—a tendency for hangers to break seems the most likely trouble here. Sparkplug life was reported to be about 8000 miles by several owners; some said that they could get longer life (more like 15,000 miles) by using an equivalent Champion plug. Running-on was a common problem with 1968s (21%), a by-product of Volvo's otherwise excellent dual-manifold emission control system. (This has been corrected for 1969 by a lower compression ratio.) Ten percent of the owners had no problems at all, other than normal maintenance and wear-tear.

Tire life ranged from 22,000 to 45,000 miles on the 1800s, 25,000 to 58,000 on the 122s; there weren't enough wear-outs on 140s to establish anything. Brake reline jobs came from 20,000 to 66,000 miles on the 122s, with the front disc brakes generally requiring attention before the rear drums—this is normal for disc/drum combinations. Seven percent of the "family" Volvos have been equipped by their owners with radial tires—some don't like the original tires' performance in the rain. (The 1800 models come with radials.) Seven percent of our owners said they had an automatic transmission and most of these made some adverse comment on it.

There wasn't a single engine overhaul reported, and the owner who had racked up 104,000 miles said, "This vehicle still has commendable compression after 104,000 miles; the head has never been off." (Our associate editor's old 544, disposed of at 97,000 miles, hadn't had its head off either—we're going to project 110,000 miles as an average life between overhauls in a Volvo.) There were only three owners who had had valve trouble, and one of those had a faulty camshaft when the car was new. Surely this durable engine is Volvo's prime claim to being an "11-year car," and it's worth noting that the engine does have to work hard most of the time. It also shows that engines don't have to be large, slow-turning and "lazy" to be durable and reliable!

In conclusion, we find that 85% of the 122S owners, 83% of the 1800S owners and 75% of the 140 owners would buy another Volvo; 5% of the 122S and 11% of the 140 owners aren't sure whether they would or not, and the rest would buy something else. Overall, 82% would buy another Volvo, a very respectable comeback potential that has, thus far in our survey series, only been exceeded by the loyalty reported by Porsche owners.

All in all, Volvo owners seem to be a practical lot who are well satisfied that they have made a wise investment in buying a Volvo and very likely will do it again.

GORDON CHITTENDEN PHOTOS

VOLVO 164

To the 144 add a British front end, luxury fittings and a 3-liter six—the result: a conservative sedan that goes

A SIX-CYLINDER Volvo has been in the mill for some time, so it was no surprise when the 164 appeared late in 1968. Details of the car weren't surprising either: it was simply a lengthened version of the 144 sedan with an inline 3-liter 6-cyl engine sharing its bore, stroke and many mechanicals with the 2-liter four of the 140 series.

It took a 3-in. increase in overall length and a 4-in. increase in wheelbase, all in the front end, to accommodate the longer engine. Volvo people say that 164 development was begun concurrently with that of the 140s; the side profile of the car seems to bear this out, looking as if the 164 were the original version rather than the stubbier-nosed 144. Viewing the 164 from the front, we are less impressed with the grille, which appears to have been lifted from late Pininfarina Wolseleys and has a diagonal slash harking back to Volvos of old.

The 6-cyl engine shares pistons, rods and valve gear with its 4-cyl counterpart but has appropriate beefing-up in areas affected by its higher output; like most modern sixes, it has a seven-bearing crankshaft. It bucks a European trend to overhead camshafts but offers instead generous displacement to get its output of 145 bhp @ 5500 rpm. Behind it is a larger clutch and a new gearbox that has a nice, short, remote-change floor lever instead of the traditional Volvo's long, slanted stick going directly into the box. A redesigned rear axle promises greater wheel bearing life and comes in ratios of 3.73:1 with manual transmission and 3.31 with the optional Borg-Warner 3-speed automatic. The axle is made by Volvo; Dana has previously supplied Volvo axles.

Suspension, front and rear, is different from that of the

140 series only in details such as spring rates. Volvo has remained conservative on the matter of wheel width, using only a 4½-in. rim width. Volvo's first power steering, a ZF unit, will be standard in the 164 for the American market.

The 164's interior makes a good initial impression and then carries through with typical Volvo livability. The high seating position is striking in a day of slinky sedans; real leather upholstery is used for the seats, and the front seats are not only sumptuously contoured but are adjustable in more ways than any we've seen. There is a level longitudinal track for fore-and-aft, a slanted and curved one that changes height and the overall seat angle, a separate seatback adjuster that allows full reclining when the adjustable head restraint is removed, and finally Volvo's exclusive lumbar-support tensioner in the lower part of the seatback. It seems almost superflous to add that these seats are comfortable! The rear bench seat, wide enough to accommodate three people rather tightly, has a fold-down armrest.

The instrument panel is identical to that of the 140 cars except for a fake wood trim strip (ugh); the traditional Volvo strip speedometer, which we now understand is intended to be read at the tip of its pointer, and the marvellously convenient pushbutton trip-odometer resetter remain, as does a radio position that's much too far from the driver. The fusebox is behind a snap-off access panel in the center, a nice convenience, and the unusual heating-ventilation dials (which are well illuminated at night) are close at hand and easy to use. Ventilation and heating are quite satisfactory, and air conditioning with its ducting reasonably well integrated into the underdash is available as an option. Vision outward in this six-window sedan is outstanding and could be improved only by elimination of the front-door vent windows.

Interiors of the 164 are available in light blue, brown or grey leather. Our white test car had the blue, a nice shade but perhaps difficult to keep clean. Door panels were two-tone; the attractive carpeting which is used on all floor areas and the rear package tray was the darker shade. There are net storage pockets in the front seatbacks for rear passengers, an ashtray in each rear door, and heat ducts to the rear compartment. There is no courtesy light, however, when the rear doors are opened.

The new engine is extremely smooth and quiet, with only

a light tappet noise from underhood to betray its presence at idle and a mild power roar on acceleration. It feels strong—much stronger than its torque or power ratings led us to expect—and revs freely, though we abided by a 5500-rpm limit in our acceleration tests because there was no official word on the redline. The engine isn't audible at freeway speeds, thanks to relatively tall gearing (for this class of car) and excessive wind noise from the front vent windows. A manual choke seems a throwback, but at least you know it's troublefree; with proper use of it the car starts and runs well from cold, though even when warm there is the same tendency we have noted in 4-cyl Volvos to stumble when starting off if enough throttle isn't used.

On a par with the excellent engine is the new gearbox. It's 100% more pleasant than the already good box in the 140s, simply because of the new shift lever location. The lever itself is an amusingly stocky affair with a huge knob, but we found it to be entirely satisfactory; the synchromesh is unbeatable, the gearbox quiet and the ratios appropriate.

And what does the 164 do with this engine and gearbox? Goes, that's what. The sly-eyed reader can't help but notice that the 164 (3 liters, 145 bhp, 4-speed box, 3.73:1 final drive, 3260 lb) clearly outperforms the MGC in this issue (3 liters, 145 bhp, 4-speed box, 3.70:1 final drive, 2915 lb). Which gets us to wondering about power ratings: the B20 engine gets 118 bhp from its 2 liters, and the 164's B30 engine is essentially 1½ B20s. So maybe the rating should be more like $118 \times 1.5 = 177$! A new kind of conservatism for Volvo?

VOLVO 164
AT A GLANCE

Price as tested	$4340
Engine	inline 6-cyl, 2979 cc, 145 bhp
Curb weight, lb	2920
Top speed, mph	110
Acceleration, 0-¼ mi, sec	17.6
Average fuel consumption, mpg	17.5

Summary: new 6-cyl model offers outstanding performance, refinement of running & accommodation at a very attractive price . . . roadholding & braking above par . . . undistinguished styling.

VOLVO 164

In ride, the 164 is little different from the 140s; tar strips and other sharp disturbances bring out some harshness, but over gentle undulations or really big bumps the 164 is soft and well controlled thanks to lots of spring travel. Rough roads upset the suspension very little, in spite of the live rear axle, but there are interior and dashboard rattles that prevent the 164 from having a rock-solid feel.

If we calculated a maneuverability-to-accommodation ratio the 164 would have to be the winner among all sedans. Its turning circle is an almost incredibly tight 31.5 ft, and its standard power steering, with a 15:1 overall ratio, removes any trace of clumsiness we have noticed in the 140s. Furthermore, this is the first power steering we've encountered that is as good as Mercedes'—it's about time somebody challenged them.

Open-road handling, too, is pleasant. The power steering imparts all the road feel you need while keeping effort low, and Volvo's suspension geometry gives a final oversteer, brought on by use of the steering wheel only, that comes in smoothly and gets the 164 around a tight turn when many sedans simply mush out. Getting off the throttle will tweak out the rear end a bit more, but at no time did we get a transition from neutral to oversteer that would be sudden enough to trip up even a moderately good driver. Even weight distribution (52% front) must have something to do with this too. Ultimate cornering speeds aren't very high—they could be greater with wider wheels and tires—but are entirely adequate for the type of car, and the overall handling characteristics are more entertaining than you'd expect in a dignified sedan.

Larger front brakes are a part of the 164 modifications, and we found them to maintain the fade resistance level of the 140 in spite of some 375 lb extra weight. Panic-stop deceleration is good, with plenty of control, and Volvo's especially good fail-safe dual hydraulic circuit (which retains part of the front brakes and one rear brake when one hydraulic circuit fails) is noteworthy. Disc brakes all around have come to be expected in cars of this class, and the 164

has them, with handbrake drums built into the rear discs.

As we have pointed out before, the trunk of this body is cavernous—and it has places for two spare tires in case the owner wants to carry two snow tires around with him. It is also adequately finished and has a light, as does the hood, for night convenience. Additional body touches are the big rubber inserts in both front and rear bumpers and a good tool kit in the trunk.

In all, the Volvo 164 is one very fine sedan for $4160. It has a tremendous trunk, roomy and luxurious interior and the great practicality of previous Volvos now combined with great refinement and, surprise of surprises, sparkling performance that requires no excuses when Volvo owners start comparing their cars with middle-priced, practical U.S. sedans powered by optional V-8 engines. Nearly everything you need is standard—power steering and brakes, whitewall tires, leather upholstery, etc. And there's no reason why Volvo's excellent engine-drivetrain durability record should not apply to the 164. The car certainly sets a new image for Volvo; the styling may not be as prestigious as that of some of its direct competitors, but the car is quite exceptional. 🔷

ROAD TEST
VOLVO 164

SCALE: 10" DIVISIONS

PRICE

Basic list...............$4160
As tested................$4340

ENGINE

Type............6 cyl inline, ohv
Bore x stroke, mm.....89.0 x 80.0
 Equivalent in........3.50 x 3.15
Displacement, cc/cu in...2979/182
Compression ratio..........9.2:1
Bhp @ rpm.........145 @ 5500
 Equivalent mph...........112
Torque @ rpm, lb-ft..163 @ 3300
 Equivalent mph...........67
Carburetion....two 1.75-in Zenith-
 Stromberg CDSE
Type fuel required......premium
Emission control....dual induction

DRIVE TRAIN

Clutch diameter, in..........9.0
Gear ratios: 4th (1.00).....3.73:1
 3rd (1.34).............5.00:1
 2nd (1.97).............7.35:1
 1st (3.14)............11.71:1
Final drive ratio.........3.73:1

CHASSIS & BODY

Body/frame...........unit steel
Brake type: disc; 10.7-in. front,
 11.6-in. rear; handbrake by aux-
 iliary drums.
 Swept area, sq in.........433
Wheels.......steel disc, 15 x 4½
Tires.....Goodyear Power Cushion
 6.85-15
Steering type..cam & roller, power
 Overall ratio...........15.7:1
 Turns, lock-to-lock.......3.7
 Turning circle, ft........31.5
Front suspension: unequal-length
 A-arms, coil springs, tube shocks,
 anti-roll bar.
Rear suspension: live axle with
 trailing arms & panhard rod, coil
 springs, tube shocks.

MAINTENANCE

Engine oil capacity, qt........6.3
Every 6000 mi: chg eng oil & filter,
 gen'l lube, cln fuel filter, cln
 plugs, adj clutch, var. op'l chks.
Every 12000 mi: chk frt-end align,
 chk compression, chg plugs.
Every 24000 mi: chg rear axle &
 trans fluid, adj auto trans, chg
 air filters.
Warranty, mo/mi.....6/unlimited

ACCOMMODATION

Seating capacity, persons...4 + 1
Seat width, front/rear.2 x 21.5/55.5
Head room, front/rear....38.0/35.5
Seat back adjustment, deg.....90
Driver comfort rating (scale of 100):
 Driver 69 in. tall..........100
 Driver 72 in. tall...........80
 Driver 75 in. tall...........70

INSTRUMENTATION

Instruments: 120-mph speedo,
 999,999 odo, 999.9 trip odo, water temp, fuel level.
Warning lights: oil pressure, alternator, brake fluid loss, high
 beam, directional signals.

EQUIPMENT

Standard: power steering & brakes,
 whitewall tires.
Optional: AM ($71) or AM/FM
 ($136) radio; automatic trans
 ($180), A/C ($354 + instal.).

GENERAL

Curb weight, lb...........2920
Test weight................3260
Weight distribution (with
 driver), front/rear, %....52/48
Wheelbase, in............106.3
Track, front/rear......53.1/53.1
Overall length............185.6
 Width..................68.3
 Height.................56.7
Ground clearance, in........7.1
Overhang, front/rear....30.2/49.1
Usable trunk space, cu ft....23.2
Fuel tank capacity, gal......15.5

CALCULATED DATA

Lb/hp (test wt).............22.5
Mph/1000 rpm (4th gear)....20.1
Engine revs/mi (60 mph)....2990
Engine speed @ 70 mph.....3480
Piston travel, ft/mi.........1575
Cu ft/ton mi..............96.3
R&T wear index.............47
R&T steering index.........1.16
Brake swept area sq in/ton....266

ROAD TEST RESULTS

ACCELERATION & COASTING

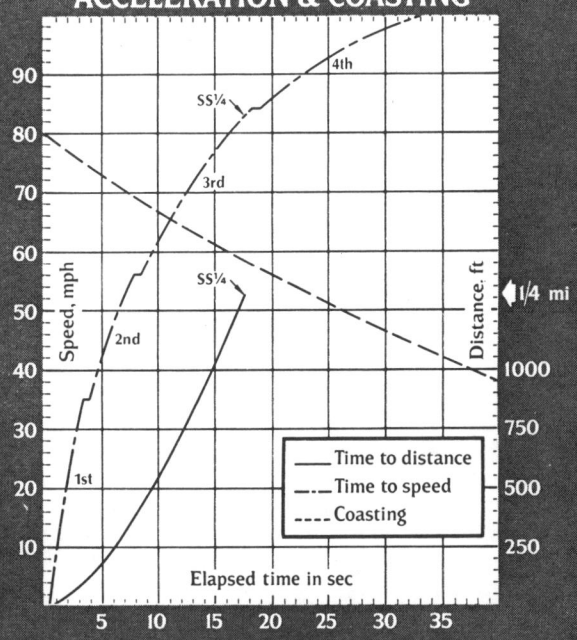

ACCELERATION

Time to distance, sec:
 0–100 ft.................3.6
 0–250 ft.................6.1
 0–500 ft.................9.5
 0–750 ft................12.3
 0–1000 ft...............14.8
 0–1320 ft (¼ mi).........17.6
Speed at end of ¼ mi, mph....83
Time to speed, sec:
 0–30 mph................2.9
 0–40 mph................4.3
 0–50 mph................6.5
 0–60 mph................9.5
 0–70 mph...............12.7
 0–80 mph...............16.3
 0–100 mph..............33.5
Passing exposure time, sec:
 To pass car going 50 mph....7.5

FUEL CONSUMPTION

Normal driving, mpg........17.5
Cruising range, mi..........271

SPEEDS IN GEARS

4th gear (5400 rpm), mph.....110
 3rd (5500)................84
 2nd (5500)................56
 1st (5500)................35

BRAKES

Panic stop from 80 mph:
 Deceleration, % g..........81
 Control............very good
Fade test: percent of increase in
 pedal effort required to main-
 tain 50%-g deceleration rate in
 six stops from 60 mph......16
Parking: hold 30% grade.....yes
Overall brake rating.....very good

SPEEDOMETER ERROR

30 mph indicated.....actual 29.6
40 mph................40.2
60 mph................60.2
80 mph................79.4
100 mph................98.0
Odometer, 10.0 mi...actual 1.042

GORDON CHITTENDEN PHOTOS

VOLVO 1800 E

The model enters its 11th year, and greatly improved performance fails to keep it competitive with newer designs

THE VOLVO 1800 coupe was first shown as a prototype in January 1960, so it qualifies nicely for Volvo's slogan "The 11-year car." Though running changes—particularly those in this new E version—have kept its performance and handling up with the times, its body is hopelessly outdated in both style and function, having never been particularly good in the first place.

E stands for *einspritzer,* we guess—for the 1800E has the Bosch electronic fuel-injection system that is being adopted by so many European manufacturers as a way to get more power, meet the American emission limits and still get clean running (see p. 21). "Brain" of the system is a box full of solid-state electronic components that lives under the dash;

ram-length intake pipes and "wilder" valve timing take advantage of the injection's precise metering to get 12 more bhp and 7 more lb-ft torque than the carbureted 2-liter Volvo engine at no increase in engine speed. Other changes for the E include jazzy styling touches headed by some cast alloy spoked wheels, flow-through ventilation system with outlets in the rear flanks and a nicer instrument panel, the beefier gearbox of the 164 and 4-wheel disc brakes as used on other Volvos. Overdrive continues to be standard equipment, and a heated rear window now is too.

Two liters and injection make the E a much stronger car than the last 1.8-liter 1800S we tested. The engine, noted for its durability rather than refinement—it's neither mechanically smooth nor quiet in the coupe—has good low-speed torque as well as the ability to pull nicely all the way to its 6500-rpm redline, and in overdrive the car will now do an honest 115 mph. The 0-60 mph and ¼-mile times are quite respectable too, putting the 1800E into the same class with such as the Alfa 1750, BMW 2500 or Mercedes 280SL. Furthermore, the engine runs cleanly without any trace of emission-control leanness symptoms and uses very little more fuel than the earlier test car.

The hefty lever on the new gearbox gives one an impression of unbreakability that is borne out by the gearbox itself; we manhandled the box unmercifully in the acceleration tests and

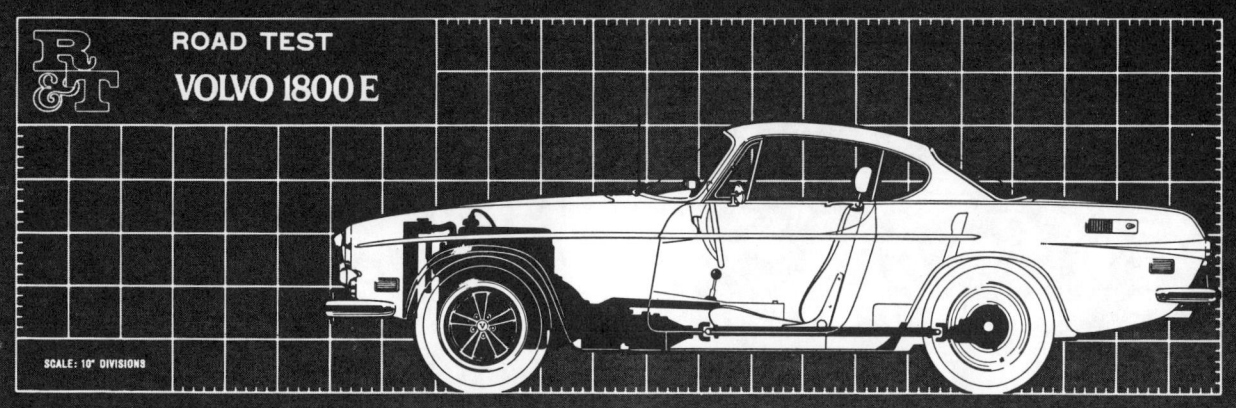

ROAD TEST
VOLVO 1800 E

SCALE: 10" DIVISIONS

PRICE

List price, east coast.......$4555
List price, west coast......$4655
Price as tested............$4655

IMPORTER

Volvo Inc.,
Rockleigh, N.J. 07647

ENGINE

Type...........4 cyl inline, ohv
Bore x stroke, mm.....89.0 x 80.0
  Equivalent in.......3.50 x 3.15
Displacement, cc/cu in...1986/121
Compression ratio.........10.5:1
Bhp @ rpm.........130 @ 6000
  Equivalent mph............133
Torque @ rpm.......130 @ 3500
  Equivalent mph.............77
Fuel injection.....Bosch electronic
Type fuel required......premium
Emission control....fuel injection

CHASSIS & BODY

Body/frame...........unit steel
Brake type: 10.6-in. disc front,
  11.6-in. disc rear, vacuum as-
  sisted.
  Swept area, sq in. 400
Wheels....cast aluminum, 15 x 5J
Tires.......Michelin XAS 165-15
Steering type......worm & roller
  Overall ratio............15.5:1
  Turns, lock-to-lock........3.25
  Turning circle, ft. 29.9
Front suspension: unequal-length
  A-arms, coil springs, tube shocks,
  anti-roll bar
Rear suspension: live axle on trail-
  ing arms with Panhard rod, coil
  springs, tube shocks

DRIVE TRAIN

Transmission.....4-speed manual
               plus overdrive
Gear ratios: o'drive (0.797)..3.42:1
  4th (1.00)..............4.30:1
  3rd (1.34)..............5.76:1
  2nd (1.97)..............8.47:1
  1st (3.14)............14.50:1
Final drive ratio..........4.30:1

ACCOMMODATION

Seating capacity, persons....2+1
Seat width, front/rear.2 x 19.5/39.5
Head room, front/rear...38.0/28.0
Seat back adjustment, degrees..10
Driver comfort rating (scale of 100):
  Driver 69 in. tall............90
  Driver 72 in. tall............85
  Driver 75 in. tall............65

INSTRUMENTATION

Instruments: 120-mph speedometer, 7000-rpm tachometer, 999,-
999 odo, 999.9 trip odo, oil press,
oil temp, water temp, fuel level,
clock.
Warning lights: brake system, overdrive, generator, high beam,
directionals, hazard flasher.

MAINTENANCE

Service intervals, mi:
  Oil change.............6000
  Filter change..........6000
  Chassis lube..........6000
  Minor tuneup.........6000
  Major tuneup........12,000
Warranty, mo/mi.....6/unlimited

GENERAL

Curb weight, lb...........2535
Test weight...............2835
Weight distribution (with
  driver), front/rear, %....51/49
Wheelbase, in............96.5
Track, front/rear......51.6/51.6
Overall length............171.3
  Width...................66.9
  Height..................50.4
Ground clearance...........6.1
Overhang, front/rear....30.2/44.6
Usable trunk space, cu ft.....8.1
Fuel tank capacity, U.S. gal...11.8

CALCULATED DATA

Lb/bhp (test weight)........21.8
Mph/1000 rpm (o'drive).....20.8
Engine revs/mi (60 mph)....2880
Engine speed @ 70 mph....3360
Piston travel, ft/mi........1510
Cu ft/ton mi (4th gear)......89.5
R&T wear index............44
R&T steering index.........0.97
Brake swept area sq in/ton....283

ROAD TEST RESULTS

ACCELERATION

Time to distance, sec:
  0–100 ft..................3.4
  0–250 ft..................6.2
  0–500 ft..................9.6
  0–750 ft.................12.3
  0–1000 ft................14.7
  0–1320 ft (¼ mi).........17.5
Speed at end of ¼ mi, mph....80
Time to speed, sec:
  0–30 mph.................3.5
  0–40 mph.................5.2
  0–50 mph.................7.1
  0–60 mph................10.1
  0–70 mph................13.1
  0–80 mph................16.8
  0–100 mph...............32.9
Passing exposure time, sec:
  To pass car going 50 mph....6.5

FUEL ECONOMY

Normal driving, mpg........20.9
Cruising range, mi..........245

SPEEDS IN GEARS

O'drive (5480 rpm)...........115
4th (6500)..................108
3rd (6500)...................80
2nd (6500)...................56
1st (6500)...................35

BRAKES

Panic stop from 80 mph:
  Deceleration rate, % g.......81
  Stopping distance, ft......310
  Control............very good
Fade test: percent increase in pedal
  effort to maintain 50%-g deceler-
  ation rate in 6 stops from 60
  mph......................25
Parking: Hold 30% grade?.....yes
Overall brake rating.....very good

SPEEDOMETER ERROR

30 mph indicated is actually...26.0
40 mph.....................35.2
60 mph.....................53.2
70 mph.....................62.3
80 mph.....................71.0
100 mph....................88.0
Odometer, 10.0 mi..........9.46

ACCELERATION & COASTING

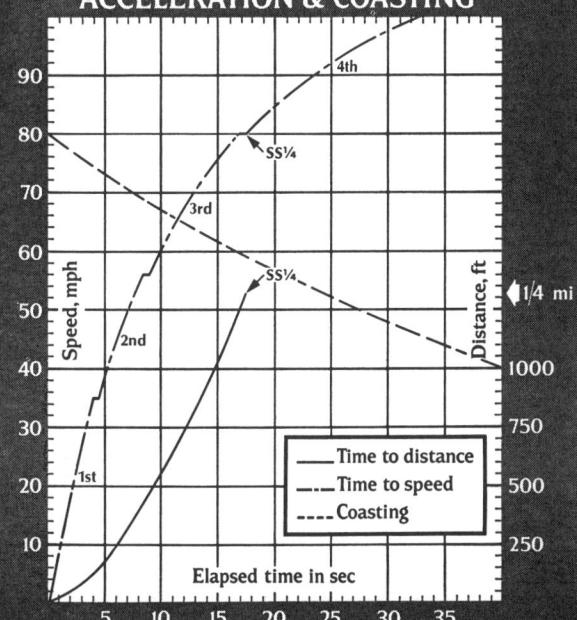

Time to distance
Time to speed
Coasting

Speed, mph
Distance, ft
¼ mi
Elapsed time in sec

Instrument layout is handsome and legible.

Tuned intake tracts, plenum chamber, injector tubes and vacuum plumbing can be seen in this view of new engine.

found it capable of taking the fastest slam shifts without a crunch. The linkage was a bit stiff on the test car but this should improve with more miles. Overdrive, the Laycock-de Normanville unit which shifts hydraulically and without lifting of the throttle foot, gives high cruising speeds with little mechanical strain although it engages jerkily.

Driving the 1800 always takes us back in time. The windshield and steering wheel (a nice new one with padded rim) are extremely close, putting the doors so far back that one has to reach across himself with his inboard arm to wind the windows; and the windowsills are so high as to make us feel as if we're sitting in a deep bathtub. The control layout isn't bad, nor are the instruments which are handsomer than ever; however, the speedometer in the test car was outrageously optimistic. Nice touches, like the good 3-point belts and the excellent seats (which, incredibly, on our test car didn't have the legally required seatback locks) with their adjustable lumbar section, are offset by poor assembly and detail design here and there—for instance, terrific wind noise caused by poorly sealing windwings or an overdrive light that's blindingly bright.

Over all kinds of roads the 1800E has a good ride though it is prone to pitching on gently undulating pavement; there is adequate vertical suspension travel so that it takes large bumps and dips in stride, and well designed linkage for the rear live axle minimizes the usual failing—primarily wheel hop—of this type of suspension. The steel-belted Michelin radials, replacing Pirelli Cinturatos used before, give noticeably increased cornering power but transmit considerable drumming to a body structure already prone to rattles.

Handling is smooth and predictable. The steering has a great feel to it that really tells the driver what's going on and to boot it's light and quick—classically good sports-car steering. In enthusiastic driving it's possible to hang out the tail by a twitch of the wheel and then keep it out by application of power—all with little danger

of overdoing anything because the car is so stable—but most drivers will be content to enjoy the pleasant response that the 1800E exhibits when merely being driven briskly. It's a good car for the novice driver, but an experienced and capable one can enjoy its handling as well.

The new disc brakes are a definite improvement over our last 1800S; panic-stop capability is up from 68% g to 81%. Fade is a moderate 25% but we really expected no fade at all in the tests; pedal effort is light but not too light and in the all-out stop the car stays in a straight line even though there is a mild tendency to wheel lockup.

To summarize, Volvo has kept the performance, handling and braking of the 1800 up to date. But in style, accommodations, refinement of running and the use of available body space, it has fallen 'way behind the times, and though it costs over $4500 it doesn't even have outstanding assembly quality to offset these drawbacks. If we were to ask for a new model from Volvo—and we sincerely hope that they aren't giving up on the sports/GT market when the 1800 runs down—we'd request a car about the same size as the 1800E but with true 2+2 seating rather than the "2+1" it now has, better noise and vibration isolation and assembly quality, and the same mechanical package underneath. *Then* we might have a Volvo GT worth $4600. ⊙

COMPARISON DATA

	Volvo 1800E	Alfa Romeo 1750 GTV	Lotus Elan S4	MGB GT (overdrive)
List price	$4655	$4681	$5133	$3628
Curb weight, lb	2535	2350	1630	2310
0–60 mph, sec	10.1	9.9	9.4	13.6
Standing ¼-mi	17.5	17.3	16.8	19.6
Speed at end	80	80	83	72
Panic stop from 80 mph, % g	81	90	87	80
Fade in 6 stops from 60 mph, %	25	nil	nil	20
R&T wear index	39	60	60	69
R&T steering index	0.97	1.26	0.79	0.93
Fuel economy, mpg	20.9	21.5	27.2	23.5

ROAD & TRACK
R&T
ROAD TEST

VOLVO 142E

A balanced set of improvements makes this the best 142 yet

ALL THE COMPONENTS are familiar; the rather boxy 142 sedan, the B20E 4-cylinder pushrod engine introduced in the 1800E last year, and the M40 gearbox with hydraulic overdrive used in the 164. Volvo has combined these to produce the 142E. Not really a Swedish supercar (there are no racing stripes), it is an improved version of the 142S with more power and more items included on the already impressive list of standard equipment.

It's a nice package. Besides the aforementioned items, the list includes 165SR-15 Pirelli tires, leather seats, electric rear window defroster, tinted glass, inertia reel seat belts and carpeting.

The B20E engine is the standard B20 engine (118 bhp @ 5800 rpm and 123 lb-ft of torque @ 3500 rpm) with Bosch electronic fuel injection, 10.5:1 compression ratio and more radical cam timing which brings it to 130 bhp @ 6000 rpm and 130 lb-ft of torque at 3500 rpm. Now the Volvo owner needn't flinch when someone questions the performance of his expensive imported sedan. The following comparisons should be helpful:

	142E	1800E	142S
Test weight, lb.	3080	2835	2812
0-60, sec.	10.5	10.1	12.3
Quarter-mile, sec.	17.5	17.5	18.5
speed at end	76	80	74

The additional 7 lb-ft of torque is not noticeable at low speeds because Volvo took advantage of the fuel injection to use more radical valve timing but it is as good at low speeds as the carbureted version. One staff member complained about having to row the car in traffic but in general we found the engine to be flexible, tractable and reasonably quiet in normal driving.

The little black box (silver, actually) which is the heart of the Bosch electronic fuel injection is located in the passenger compartment under the right front seat, thus keeping it away from the damaging heat of the engine compartment.

Freeway driving is the forte of the 142E, providing relaxed (both car and passengers) cruising at and above freeway speeds. The reason for this is the Laycock-de Normanville overdrive unit which changes the busy 4.33:1 final drive ratio to a more relaxed 3.42:1. A stalk on the steering column controls the unit and although it isn't necessary to use the clutch when going into overdrive, we found using it eliminated a clunk when it engaged.

Handling is typically Volvo, that is, lots of body lean and final oversteer. One feels secure in a Volvo, even during hard cornering, despite what you see from the outside as a spectator. Outside front tire ready to tuck under, inside rear wheel lifting off the pavement and a horrendous amount of body lean makes for scary viewing. The 142E gave a good account of itself on the skid pad with a speed of 31.2 mph and a lateral acceleration figure of 0.649g, the Pirelli tires doing their job despite the drama.

COMPARISON DATA

	Volvo 142E	Alfa Romeo 1750 Berlina	BMW 2002
List price, incl. prep	$4080	$3680	$3426
Curb weight, lb	2696	2484	2210
0-60 mph, sec	10.5	11.0	10.4
Standing ¼-mi	17.5	17.9	17.6
Speed at end	76	76	78
Stopping distance from 80 mph, ft	306	n.a.	380
Fade in 6 stops from 60 mph, %	20	nil	21
Cornering capability, % g	0.649	n.a.	n.a.
Fuel economy, mpg	23.7	20.4	25.0

VOLVO 142E

Another Volvo trait, the small turning circle, is well appreciated. The 16-ft car will turn curb-to-curb in a bit over 30 ft.

The interior, as with all the 140 series Volvos, is comfortable. All but one of the staff liked the seats with their adjustable backrests and soft-to-firm settings for lumbar support.

Seat belts deserve a mention here too. The idea, 3-point belts on inertia reels, is good and superior to most systems in common use but the examples in the two cars we drove were frustrating. In both cars the passenger belts were excellent, allowing slow forward movement but locking at the slightest tug. The driver's belts, however, seemed to have minds of their own. The first allowed the driver to move anywhere he wanted; unfortunately it would allow him to go

through the windshield before it locked. The other was so sensitive it refused to be unrolled at all.

Our usual gripes about not having enough instruments, especially a tachometer, apply here. Also, the red light which tells the driver when overdrive is engaged is an unfortunate choice. It is a bad habit, in a car which relies on warning lights, to get used to a red light glowing on the dashboard.

The 142E renewed the Volvo speedometer controversy. It is a ribbon design with a slash cut in the red indicator to provide a pointer. The question of which portion of the ribbon to read comes up every time we test a Volvo. Reading at the pointer, which seems most logical, has the speeds five mph fast on the entire scale. Reading behind the slash, where the indicator is full width, gives spot-on speedometer readings.

A tall 57 inches, the car has excellent vision in all directions. Location of the fuse box in the center of the dashboard beneath the heater controls makes for easy service.

A Blaupunkt AM-FM stereo radio was the only extra on our test car. It is an expensive option but made the long

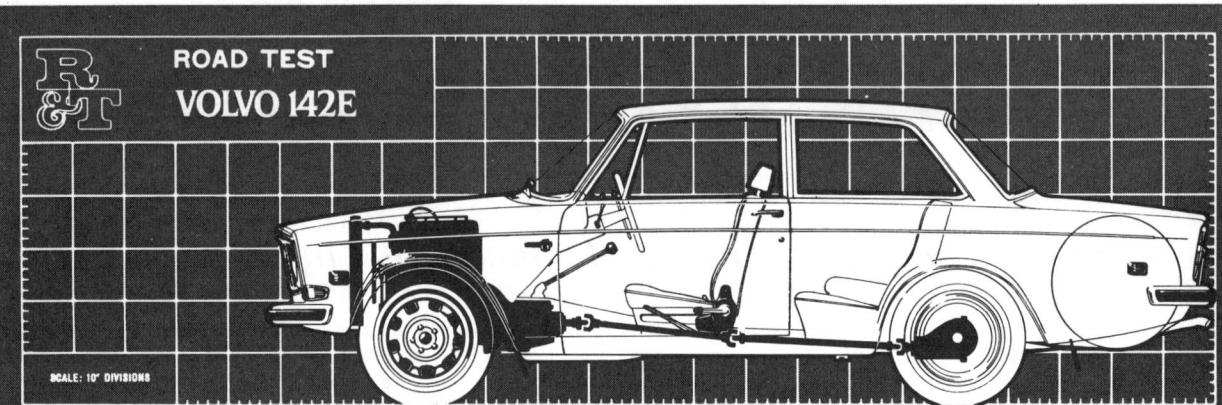

ROAD TEST
VOLVO 142E

SCALE: 10" DIVISIONS

PRICE

List price, east & Gulf coasts..$3895
List price, west coast.......$3990
Price as tested, west coast..$4272
Price as tested includes standard equipment (overdrive, radial tires, front & rear disc brakes with vacuum assist, rear window defroster), Blaupunkt AM-FM stereo radio ($192), dealer prep ($90).

IMPORTER

Volvo, Inc.
Rockleigh, New Jersey 07647

ENGINE

Type................ohv inline 4
Bore x stroke, mm.....89.0 x 80.0
 Equivalent in.......3.50 x 3.15
Displacement, cc/cu in...1986/121
Compression ratio........10.5:1
Bhp @ rpm.........130 @ 6000
 Equivalent mph (4th).......106
Torque @ rpm, lb-ft..133 @ 3500
 Equivalent mph............64
Fuel injection.....Bosch electronic
Type fuel required......premium
Emission control....fuel injection, engine mods

DRIVE TRAIN

Transmission.....4-speed manual with overdrive
Gear ratios: OD (0.79).....3.42:1
 4th (1.00)............4.33:1
 3rd (1.36)............5.88:1
 2nd (1.99)............8.61:1
 1st (3.13)...........13.55:1
Final drive ratio...........4.33:1

CHASSIS & BODY

Layout.....front engine/rear drive
Body/frame............unit steel
Brake type: 10.7-in disc front, 11.6-in disc rear; vacuum assisted
 Swept area, sq in.........400
Wheels.........steel disc, 15 x 5
Tires...Pirelli Cinturato 165 SR-15
Steering type........cam & roller
 Overall ratio............17.5:1
 Turns, lock-to-lock.........4.0
 Turning circle, ft.........30.4
Front suspension: unequal-length A-arms, coil springs, tube shocks, anti-roll bar
Rear suspension: live axle on trailing arms & Panhard rod; coil springs, tube shocks

ACCOMMODATION

Seating capacity, persons....4+1
Seat width, front/rear 21.5 x 2/54.5
Head room, front/rear..38.0/36.0
Seat back adjustment, degrees..90

INSTRUMENTATION

Instruments: 120-mph speedometer, 99,999 odometer, 999.9 trip odometer; fuel level, coolant temperature
Warning lights: oil pressure, alternator, brake-on, directionals, high beam, overdrive

MAINTENANCE

Service intervals, mi:
 Oil change..............6000
 Filter change...........6000
 Chassis lube............none
 Minor tuneup...........6000
 Major tuneup.........12,000
Warranty, mo/mi.....6/unlimited

GENERAL

Curb weight, lb...........2695
Test weight..............3070
Weight distribution (with driver), front/rear, %....52/48
Wheelbase, in...........103.1
Track, front/rear......53.1/53.1
Overall length..........182.7
 Width................68.3
 Height...............57.7
Ground clearance.........7.1
Overhang, front/rear...31.2/49.1
Usable trunk space, cu ft.....23.2
Fuel tank capacity, U.S. gal..15.5

CALCULATED DATA

Lb/bhp (test weight)........23.6
Mph/1000 rpm (4th gear).....17.0
Engine revs/mi (60 mph)....3540
Piston travel, ft/mi.......1865
R & T steering index........1.22
Brake swept area sq in/ton....260

RELIABILITY

From R&T Owner Surveys the average number of trouble areas for all models surveyed is 10.6. As owners of earlier model Volvos reported 10 trouble areas, we expect the reliability of the Volvo 142E to be average.

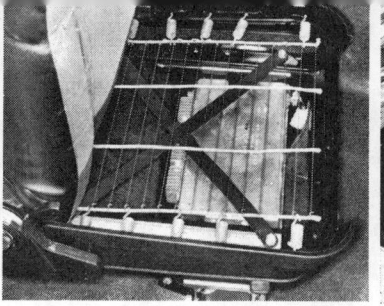

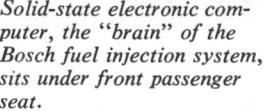

Solid-state electronic computer, the "brain" of the Bosch fuel injection system, sits under front passenger seat.

freeway trips at which this car excels more enjoyable.

In all, we found the 142E to be an improved version of the 142S, rather than a sporting sedan. If one doesn't mind shifting often in traffic, the comfort, relaxed cruising and fuel economy make it worth consideration. ▣

ROAD TEST RESULTS

ACCELERATION

Time to distance, sec:
0–100 ft	3.4
0–250 ft	6.4
0–500 ft	9.5
0–750 ft	12.4
0–1000 ft	14.9
0–1320 ft (¼ mi)	17.5
Speed at end of ¼ mi, mph	76

Time to speed, sec:
0–30 mph	3.6
0–40 mph	5.2
0–50 mph	7.2
0–60 mph	10.5
0–70 mph	14.7
0–80 mph	20.0
0–90 mph	27.1

Passing exposure time, sec:
To pass car going 50 mph....7.2

FUEL CONSUMPTION

Normal driving, mpg	23.7
Cruising range, mi	360

SPEEDS IN GEARS

O.D. (4720 rpm)	106
4th (6000)	106
3rd (6300)	82
2nd (6300)	58
1st (6300)	38

BRAKES

Panic stop from 80 mph:
Max. deceleration rate, % g	81
Stopping distance, ft	306
Control	good
Pedal effort for 50%-g stop, lb	25

Fade test: percent increase in pedal effort to maintain 50%-g deceleration rate in 6 stops from 60 mph....20
Parking: Hold 30% grade?.....yes
Overall brake rating.....very good

HANDLING

Speed on 100-ft radius, mph	31.2
Lateral acceleration, % g	0.649

SPEEDOMETER ERROR

30 mph indicated is actually	25.0
40 mph	35.0
50 mph	45.0
60 mph	54.0
70 mph	64.0
80 mph	73.0
100 mph	93.0
Odometer, 10.0 mi	9.9

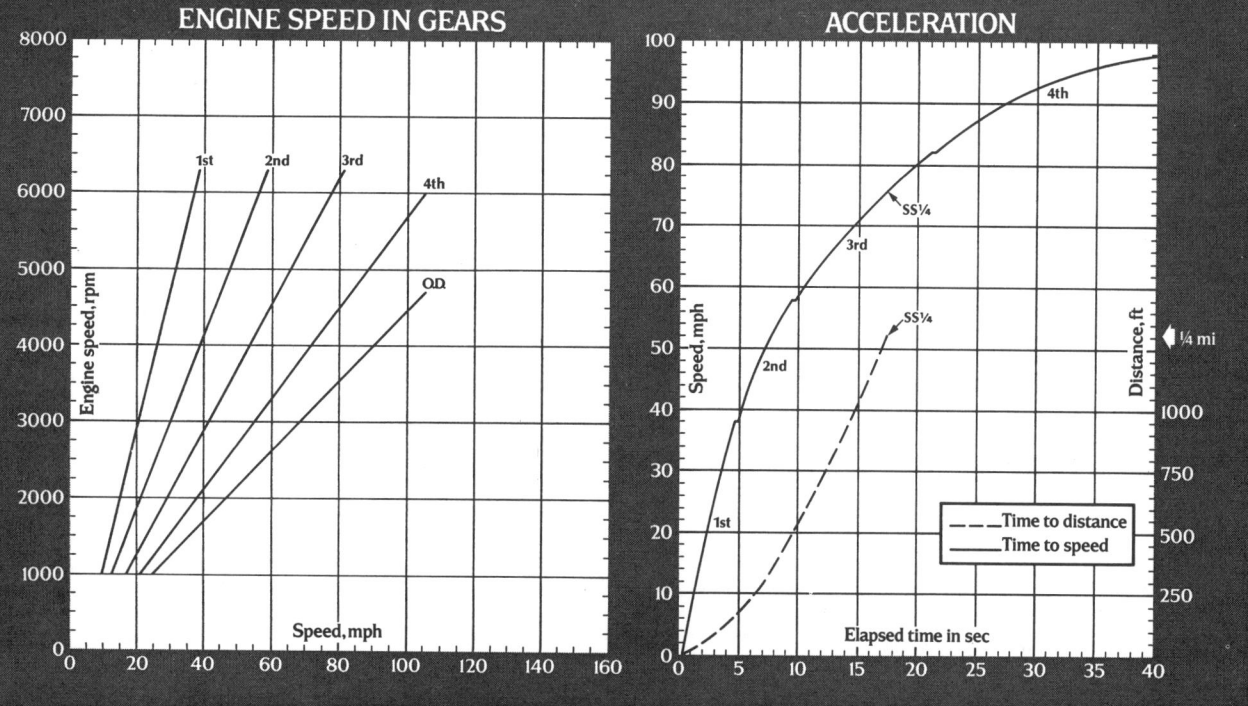

ENGINE SPEED IN GEARS

ACCELERATION

VOLVO

*Parts sources and tips
(plus two examples)
for the Volvo owner
who wants more than
transportation. The
first of a series*

BY MIKE ANSON

Volvo B20 engine with factory tuning kit. Beneath the carburetors is tube exhaust.

GENERAL		SUSPENSION			ENGINE		
Workshop and Repair Manuals	Tuning Books and Catalogs	Wheels	Springs and Shock Absorbers	Anti-roll Bars	Tuning Kits	Carburetors and Manifolds	Camshafts
Volvo Owners Manual $15.00 (V)	Volvo Performance Tuning (V)	Slotted steel disc, 15 x 4½-in. 66820; solid steel disc 15 x 4½-in. 657838 (V)	Progressive coil springs to fit 140 series sedans; front 552105, rear 677252 (V)	Rear to fit 140 series $24.95, 122S & 1800 $29.50, 544 $24.95 and 164 $29.50 (IP)	B-18 engine 419398, B-20 engine 419881 $575.00 (V)	Carburetors 237435 and intake manifolds 419883 and 419884 (V)	B-18 and B-20; camshaft 419258(D) and valve springs 418737 (V)
Volvo Owners Handbook—Floyd Clymer $5.00 (a)	Volvo Competition Service Catalog (V)	Slotted steel disc widened to 6 or 7-in. 14 x 6 or 7-in. $19.95 15 x 6 or 7-in. $20.95 (IP)	Competition coil springs; shorter coil with larger diameter wire $24.95 pr. (IP)	Front to fit 140 series, ¾-in. diameter 12019 $22.95; ⅞-in. diameter 14019 $23.50; rear, 1-in. diameter 1637 $24.95; 1⅛-in. diameter 1837 $24.95 Addco, Inc., Watertower Road, Lake Park, Fla. 33403	High Power Kit— 130 hp, B-18 or B-20 $295.00 (IP)	Weber Carburetor Kit—MC2044 includes 2 Weber carbs with manifold and linkage $258.00 Ernie McAfee, 4091 Redwood, Los Angeles, Calif. 90066 Weber carburetor kits also available from BAP/GEON dealers (b)	Ed Iskenderian Racing Cams, 16020 S. Broadway, Gardena, Calif. 90247 VV-71 street use, VV-81 competition, regrind $50.00, outright $85.00 (also available from IP)
Glenn's Tune-up & Repair Guide for Volvo $4.50 (a)	Import Parts Dist. Co., Parts, Accessories and Speed Equipment for Volvo (IP)	American Racing Co. 15 x 6-in. G.T.P. style alloy wheel $99.50 each (b)	Shock absorbers; front 552103, rear 552104 (V)				Racer Brown Cams, 9270 Borden, Sun Valley, Calif. FGV-street $40.00, CTV-4-street or track $40.00, V290X-competition $40.00
Chilton's Repair Guide for Volvo $4.95 (a)		Minilite magnesium 15 x 6-in. C115-5Q3 $90.15 140 series use C1555-5L3 $90.15 (b)	Spax adjustable $19.95 each (various types available; see catalog) Traction bars $44.95 (IP)				Weber Camshafts, 310 S. Center, Santa Ana, Calif. 92703 19-A street or slalom $45.00, ST-15 competition $50.00, ST-12 super competition $50.00
Handbook of Foreign Car Carburetors— Floyd Clymer $5.00 (a)		Fenton cast aluminum 15 x 6-in. CW6343; with spokes CW8443 (b)	Koni to fit 140 series; front 80A-1319, rear 80C-1932. 122 series; front 80A-1319, rear 80A-1320 (b)				NOTE: Most major cam grinders will regrind imported car camshafts.
Glenn's Foreign Car Repair (includes Volvo) $17.50 (a)		Ford or Plymouth 14 x 6-in. will fit up to 140 series					

(a) available from: Autobooks   Classic Motorbooks   Robert Bentley, Inc.   (V) available from local Volvo dealer or   Volvo Distributing, Inc.
                 2900 R. Magnolia   3844 Thomas Ave. South   872 Mass .Ave.   Rockleigh, New Jersey 07647
                 Burbank 12, Calif.   Minneapolis, Minnesota 55410   Cambridge, Mass. 02139

(b) available from imported car parts stores or speciality shops

VOLVO'S REPUTATION AS sturdy, dependable, weather-proof transportation is seldom challenged. Performance hasn't been as important. The factory and some of the more enthusiastic owners foray into rallying and sedan racing on occasion, but the focus has been on years, not seconds. Despite this—or perhaps because a sturdy car lends itself to having more performance extracted from it, a group of speciality shops making and selling Volvo equipment does exist.

The factory is getting into the act, too. Volvo's Competition Department in Göteborg is expanding its program and offers carburetors, camshafts, cylinder heads, driving seats, rollbars, steering wheels and lightweight body parts. An added feature in the factory's help is that installation of the pieces doesn't void the warranty.

This is not a how-to-do-it article. Rather, the purpose is to list what is available and describe what has been done by others. Each owner can then decide what modifications he wants or needs or both.

Improved 142S with special grille, driving lights and P1800 wheels.

Optional instrument cluster was only interior modification.

Volvos in General

THE TWO most common Volvo engines, and the ones we deal with here, are the B18 and B20. Both are inline 4-cylinder pushrod engines with a reputation for longevity. Displacing 1778 cc, with a bore and stroke of 84.1 x 80.0 mm, the B18 was introduced on the 1962 Volvos. Offered in the sedan with 90 bhp, this same engine in the P1800, with compression raised from 8.5:1 to 9.5:1, was rated at 100 bhp. Highest output version of the B18 was in the 1800S, rated at 115 bhp. Increased compression (10.0:1) and a wilder camshaft accounted for the 15 bhp increase.

The B20 engine, essentially a bored-out B18, was first offered on 1969 Volvos. Stroke remained the same at 80.0,

		DRIVE TRAIN		BRAKES	ACCESSORIES		
Heads	Exhaust Systems	Clutch and Flywheel	Differential		Seats	Roll Bars	Instruments
B-18 complete with valves and springs 419392, B-20 419882 (V)	Branched exhaust manifold 419381, "rallye" exhaust pipe 552211 (V)	B-18 Flywheel 419392, B-20 same (V)	4:88:1 Anti-spin 538445; Anti-spin unit 384398, 4:88:1 gearset 525880 (V)	Competition brake pads; front 275894, rear 275895 (V)	Dished driving seat 552301 and base 552303 (V)	B & B Motors, Burnt Hills, N.Y. 12027	Instrument cluster 552701 includes speedometer, 8000 rpm tachometer, oil pressure, coolant temp. and fuel level (V)
High compression cylinder head (11.7:1) $195.00 exchange (IP)	Four-tube equal length exhaust headers $89.50; exhaust kit, 2 mufflers and cross-over pipe $41.95 (IP) Four-tube exhaust header kits $30.95 Autopower, 3163 Adams, San Diego, Calif. Four-tube exhaust headers for 1967 and earlier B18 $69.95 Autoworld, Inc., 701 N. Keyser, Scranton, Pa. Abarth exhaust system, 140-VOL142 $89.50, 140 deluxe-VOL144 $114.50 and others (b) Peco exhaust system 122S-TBB17 $31.00, P1800S-TBB16 $26.00 (b)	Lightened flywheel $24.95 exchange, H.D. clutch cover $24.95 exchange, H.D. clutch plate $19.95, aluminum flywheel $145.00, diaphragm clutch for 140 series $35.95 (IP)	Power-Lock limited-slip for 544, 122S and P1800 $70.00, available soon for 142 and 144 (IP)	Cali-Loc brake pads $6.00 set exchange, shoes $15.95 set. Competition Mione pads $60.00 set (IP)	Racing bucket seat (sit-up style) $99.50, street bucket seat $22.95 Autoworld, Inc., 701 N. Keyser, Scranton, Pa.	(available from Volvo at a later date)	Smiths oil pressure and coolant temp $24.95, Smiths tach $44.95 (IP) (local speciality shops and mail order houses such as Autoworld, MG Mitten, Vilem B. Haan and Wilco also stock such items)

Volvo Western Distributing Co.
1955 190th St.
Torrance, Calif. 90509

Volvo Southwest, Inc.
3303 West 12th St.
Houston, Texas 77008

(IP) Import Parts Distributing Co.
2362 N.E. Broadway
Portland, Oregon 97232

other sources: Unex Products Corp.
37 West 65th St.
New York, N. Y. 10023

K. N. Rudd
41 High Street
Worthing, Sussex, England

VOLVO

but the bore was increased from 84.1 to 89.0 mm. Power rating increased only to 118 bhp.

The simplest method of increasing the performance of the lower-horsepower version B18 or increasing the power of the B20 (which uses a milder cam) is to backfit parts and bring the engine to 1800S specifications. In the case of the B20, count the additional displacement as a bonus.

Improvements and changes don't have to stop here.

Volvo engines are sturdy (there is that word again). In fact, the lower end of the Volvo engine is good for sustained running at 6000 rpm with occasional trips to 6500. It does need help breathing and both the factory and the aftermarket can provide that help.

Carburetion is easy; there are two choices. 1) Modify the stock carburetors. Recommended changes are included in the *Volvo Tuning Manual* available from Volvo dealers. 2) Buy a Weber system from a BAP/Geon dealer or McAfee in Los Angeles, or the factory high performance system with Solex carburetors.

High-performance cylinder heads are available from speciality shops or from Volvo. The stock heads on most B18 engines can be improved by milling 0.020 in., which brings the compression to 10.5:1, and installing larger valves (optional) and stronger valve springs.

A set of exhaust headers connected to an efficient muffler system completes the breathing modifications.

It is important to note that with engine modifications one thing always leads to another. All the carburetors you can bolt on won't improve a thing and will, in fact, hurt performance unless you also improve the camshaft and exhaust system. Modifying the stock carburetors does not require these other operations.

Replacing the stock clutch assembly with a high performance unit is a must. Volvo makes a special clutch for this purpose, or it is possible to have the stock clutch assembly modified. Without a proper clutch to transmit the added power and torque all the engine modifications are for naught.

Volvo offers a variety of transmission gear ratios, but only the racer will need to change the stock gearbox. A limited-slip differential is definitely recommended. Volvos tend to lift the inside rear wheel during hard corners and without a limited slip, the wheel remaining on the ground has no drive. No fun at all.

Anti-roll bars and adjustable shock absorbers are an easy way to better handling. Both the factory and the specialty shops offer handling packages with special springs and lowering kits for those with the inclination.

Owners of older Volvos (pre-disc brake) interested more in function than form can fit 5½- or 6-in.-wide wheels simply by bolting on a set of 1956-1965 Ford or Plymouth station wagon rims. Even the lug bolts are the same.

A full complement of gauges is a necessity. Any oil pressure gauge used with a Volvo must have a scale of at least 100 due to the Volvo's high oil pressure. A gauge with a lower scale will be blown as soon as the engine is started. A selection of gauges from a sports car shop, or the factory dash panel with speedo, tach and oil pressure will let the

One man's hobby. Slalom Volvo cost $600.

driver know what is happening in the engine compartment.

Examples

ONE EXAMPLE of an Improved Performance Volvo is a new 142S driven by Bob Sinclair, President of Volvo Western Distributing Co.

This car got the entire treatment. First, tuning kit 419881 which includes cylinder head, carburetors, camshaft kit, tubing exhaust headers, and necessary brackets and linkage was installed.

Next, a set of "rally" springs (numbers are listed in the chart) which raised the car 2 in. Heavy duty non-adjustable shocks were added to improve the handling. A set of Volvo competition brake pads was installed and a 4.33:1 limited-slip differential was also fitted.

Pirelli Cinturato tires, size 185SR-15, were mounted on 1800S wheels. A new grille with driving lights completed the exterior changes.

Inside, a new instrument cluster with 8000-rpm tachometer, speedo, fuel level and oil pressure gauge was added.

The entire operation was a bolt-on affair. Cost for the engine tuning kit is high ($575) but the various parts are available separately.

This improved performance Volvo was not difficult to drive, nor was the engine unduly fussy. Prolonged idling would foul the plugs, but a blast to the redline would bring everything back into tune.

This car does serve as an example of what often happens after an engine modification; the stock clutch won't hold the engine and as a result the performance suffers. The rest of the Volvo driveline is so tough it doesn't need beefing-up but without the heavy duty clutch the engine couldn't be used to potential.

Only the Volvo purists could spot the differences, so we had lots of fun with this good handling and powerful Volvo. To follow this example is expensive, but by choosing the right assortment of parts a Volvo owner can improve his car at significantly lower cost.

A slalom winning 544 is our second example. Leo Baker of Tucson, Arizona has $600 invested in a Volvo which can still be driven on the street. The car is gutted, with a fiberglass bucket seat (passenger seat removed for racing) and roll bar the only interior appointments.

Engine modifications are: an Isky camshaft, do-it-yourself headers from a mail-order company, electric fuel pump and air horns for the carburetors. The head has never been removed during the 85,000 miles (so far) and the engine still sounds and feels strong.

The post-war Fordish 544 squats on fat tires, with the fenders bulged out to accommodate them.

Baker cut one coil from the front springs to lower the car 3 in. at the front. A Porta-power, a sandbag and a 2 x 4 made the neat looking bulges in the fenders so the 15 x 6-in. Ford wheels and tires would clear. The front wheels are set with 2-degree negative camber. No one sells a larger diameter front anti-roll bar, so Baker, using the old adage if one is good, two is better, uses two anti-roll bars, on the front.

At the rear, the springs were heated to collapse the coils and drop the rear 2 in. The Porta-power again provided the necessary clearance. Incidentally, more clearance is needed on the right side of the car because of the Panhard rod. The methods sound crude, but they are cheap and effective.

Yes, the budget hobby car does exist.

Where to Find it

A LIST of suppliers follows the chart. If something you need is not listed, try anyway; these people are adding new products constantly. Addresses of Volvo distributors are included in case the local dealer gets that puzzled look when you ask for the high-performance parts. Prices on the Volvo factory parts will be available soon.

VOLVO 164E

Sweden's fuel-injected luxury transportation machine

THE FUEL INJECTION is new and there have been some minor trim changes, but the 164E is basically an improved version of Volvo's luxury sedan which we tested shortly after it was introduced in late 1968.

There is some good news and some not-so-good news regarding the engine, but fortunately the good outweighs the bad in the emissions/performance teeter-totter. The good news is that Bosch electronic fuel injection has been added to the 2979-cc, 6-cyl engine, now called the B30F. And the experience of having a similar system on two 4-cyl cars previously (1800E and 142E) has resulted in modifications and improvements for the fuel injection used on the 164E. An improved cold-start system (fully automatic; the driver doesn't even touch the accelerator when starting up) is one.

Injection timing is also changed; the injectors now squirt on closed intake valves, a trick Volvo found to reduce un-

COMPARISON DATA

	Volvo 164E	BMW Bavaria	Mercedes 250
List price, incl. prep	$5080*	$5485	$7026*
Curb weight, lb	3040	3170	3150
0–60 mph, sec	12.0*	9.3	13.6*
Standing ¼-mi, sec	18.8*	16.8	19.0*
Speed at end	76*	82	72*
Stopping distance from 80 mph ft	397	300	300
Fade in 6 stops from 60 mph, %	12	nil	nil
Cornering capability, g	0.690	0.726	n.a
Fuel economy, mpg	17.7*	18.0	16.0*

*automatic transmission

VOLVO 164E

burned hydrocarbon emissions. A new distributor cam was required for this—making the distributors from fuel injected and carbureted Volvo engines no longer interchangeable.

The not-so-good news is that the compression ratio of the 164E, like that of all Volvos this year, has been reduced from 9.2:1 to 8.7:1 to allow the engine to run on 91—octane fuel. This was achieved by raising the height of the cylinder head, which means an enterprising enthusiast wanting more power and better fuel economy could remove the head and have it milled to raise the compression ratio. The results of low compression have been offset somewhat by the fuel injection. For one, the fuel consumption of the 164E with automatic transmission equaled that of the carbureted 164 with 4-speed manual transmission and 9.2:1 compression ratio—17.7 mpg. The rated horsepower (SAE net) is now

springs in front, a live axle with longitudinal and lateral linkage and coil springs in the rear. The brakes are disc, front and rear. The 10.7-in. front discs are now vented for better cooling and more resistance to fade. Although brake fade has not been a problem with Volvos lately, the vented discs do reduce it: in our previous 164 road test the brakes faded 16% in our 6 stops from 60 mph, but the 164E's brakes faded only 12%. Stopping distance from 80 mph was not as impressive, however; the 164E needed a most unsatisfactory 397 feet to stop. This may be the fault of the smallish 165-15 tires, which the brakes can easily overpower and lock up. The 3040-lb car should have larger tires, and we would suggest 185-15 on the 5½-in. rims that are standard.

Outside, the changes are minor. The door handles are now recessed and that's about the only way to tell a new Volvo from a pre-1972 model. Most people are used to the boxy, rather tall shape by now and it's pleasing enough to look at—if Volvo didn't overdo it with trim, identification badges and vinyl tops. One benefit of the Volvo's nondescript

Electronic box for fuel injection is under the passenger seat.

136 @ 5800 rpm and the torque is 154 lb-ft @ 2500 rpm; the original carbureted 164 did 130 bhp net and 152 lb-ft.

Our test car was equipped with an automatic transmission, a revised 3-speed Borg-Warner unit. Judging from the list of changes in the gearbox, it should be a type 65 but is listed in the specifications as the Borg-Warner type 35. In any case the new gearbox has a wider front band with upgraded friction material and a new governor for improved shift patterns. The rear pump has been deleted (no more push starts) and the front pump enlarged. The torque converter ratio is now 2.3:1 versus last year's 2.1:1.

Suspension remains the same: unequal A-arms and coil

appearance is that it is less likely to be noticed by the local traffic enforcement agency if the speed happens to creep above the posted limit—which is easy enough in the 164E.

It is on the freeways that the 164E, and indeed most Volvos, excel. It offers relaxed cruising at freeway speeds with just a trace of tappet noise and a bit of wind whistle. Too, the people package is so nicely done that the pleasure of owning a Volvo is in using it, not simply looking at it. A plush interior with leather seats, carpeting and nice trim really makes the 6-cyl sedan a truly luxurious car and belies its modest exterior.

In addition to adjustments forward and backward and for

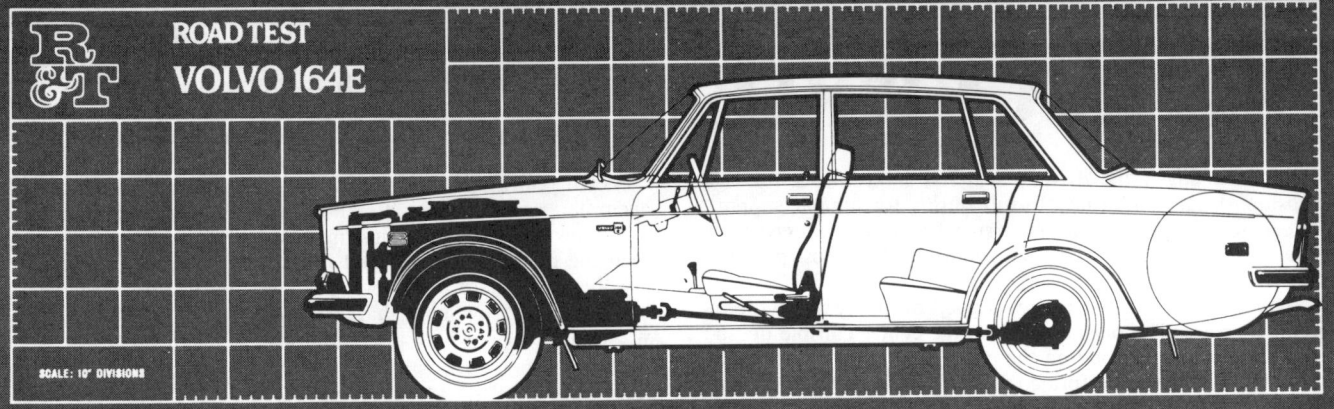

R&T ROAD TEST
VOLVO 164E

SCALE: 10" DIVISIONS

PRICE

List price, west coast......$5050
Price as tested, west coast..$5080
  Price as tested includes standard equipment (automatic transmission, radial tires, leather upholstery, rear window defroster), dealer prep ($30)

IMPORTER

Volvo Inc., Rockleigh, N.J. 07647

ENGINE

Type................ohv inline 6
Bore x stroke, mm.....89.0 x 80.0
  Equivalent in.......3.50 x 3.15
Displacement, cc/cu in...2979/182
Compression ratio..........8.7:1
Bhp @ rpm..........138 @ 5800
  Equivalent mph.............115
Torque @ rpm, lb-ft..154 @ 2500
  Equivalent mph..............51
Fuel injection....Bosch electronic
Type fuel required: regular, 91-oct
Emission control....fuel injection, engine mods

DRIVE TRAIN

Transmission...automatic; torque converter with 3-sp planetary gearbox
Gear ratios: 3rd (1.00).....3.31:1
  2nd (1.45)..............4.80:1
  1st (2.39)..............7.90:1
  (2.39 x 2.3)...........18.20:1
Final drive ratio.........3.31:1

CHASSIS & BODY

Layout....front engine/rear drive
Body/frame.............unit steel
Brake type.......10.7-in vented disc front, 11.6-in disc rear; vacuum assisted
  Swept area, sq in.........433
Wheels.....steel disc, 15 x 5½J
Tires...Pirelli Cinturato 165 SR-15
Steering type...recirculating ball, power assisted
  Overall ratio............15.7:1
  Turns, lock-to-lock........3.7
  Turning circle, ft.........32.8
Front suspension: unequal-length A-arms, coil springs, tube shocks, anti-roll bar
Rear suspension: live axle with trailing arms & Panhard rod, coil springs, tube shocks

ACCOMMODATION

Seating capacity, persons....4+1
Seat width,
  front/rear.......2 x 21.5/55.5
Head room, front/rear..39.0/35.5
Seat back adjustment, degrees..90

INSTRUMENTATION

Instruments: 0–120 mph speedo, 99,999 odo, 999.9 trip odo, coolant temp, fuel level, clock
Warning lights: oil pressure, alternator, brake fluid loss, brake-on, high beam, directionals, hazard flasher, seat belts

MAINTENANCE

Service intervals, mi:
  Oil change................6000
  Filter change............6000
  Chassis lube.............none
  Minor tuneup...........6000
  Major tuneup.........12,000
  Warranty, mo/mi.....6/unlimited

GENERAL

Curb weight, lb.........3040
Test weight............3420
Weight distribution (with driver), front/rear, %....54/46
Wheelbase, in............107.1
Track, front/rear......53.2/53.2
Overall length...........185.6
  Width..................68.1
  Height.................56.7
Ground clearance..........7.1
Overhang, front/rear....29.8/48.7
Usable trunk space, cu ft....23.2
Fuel tank capacity, U.S. gal...15.3

CALCULATED DATA

Lb/bhp (test weight).........24.8
Mph/1000 rpm (3rd gear)....20.1
Engine revs/mi (60 mph)....2980
Piston travel, ft/mi........1560
R & T steering index........1.22
Brake swept area sq in/ton...253

RELIABILITY

From R&T Owner Surveys the average number of trouble areas for all models surveyed is 11. As owners of earlier model Volvos reported 10 trouble areas, we expect the reliability of the Volvo 164E to be average.

ROAD TEST RESULTS

ACCELERATION

Time to distance, sec:
  0–100 ft...................4.5
  0–250 ft...................7.5
  0–500 ft..................10.7
  0–750 ft..................13.5
  0–1000 ft.................16.1
  0–1320 ft (¼ mi)..........18.8
Speed at end of ¼ mi, mph...76.0
Time to speed, sec:
  0–30 mph...................4.9
  0–40 mph...................6.7
  0–50 mph...................9.0
  0-60 mph..................12.0
  0–70 mph..................15.8
  0–80 mph..................21.1
  0–90 mph..................29.1
Passing exposure time, sec:
  To pass car going 50 mph...6.0

FUEL CONSUMPTION

Normal driving, mpg........17.7
Cruising range, mi..........270

SPEEDS IN GEARS

3rd gear (5800 rpm).......115
2nd (6000)................89
1st (6000)................57

BRAKES

Panic stop from 80 mph:
  Max. deceleration rate, % g..84
  Stopping distance, ft......397
  Control..........very good
Pedal effort for 50%-g stop, lb..25
Fade test: percent increase in pedal effort to maintain 50%-g deceleration rate in 6 stops from 60 mph..................12
Parking: Hold 30% grade?....yes
Overall brake rating.........fair

HANDLING

Speed on 100-ft radius, mph..32.1
Lateral acceleration, g.......0.690

SPEEDOMETER ERROR

30 mph indicated is actually..31.0
40 mph....................40.5
50 mph....................50.0
60 mph....................59.0
70 mph....................68.5
80 mph....................78.0
Odometer, 10.0 mi......9.8

ACCELERATION

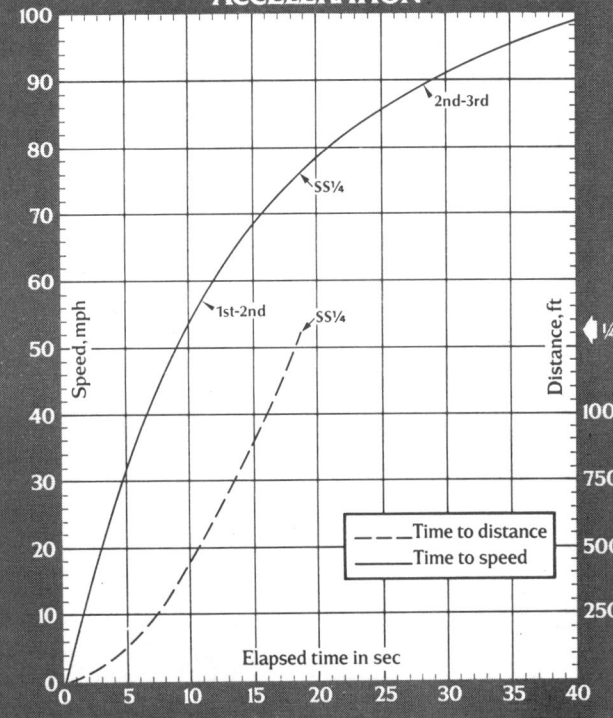

Legend: - - - - Time to distance — Time to speed

Elapsed time in sec

VOLVO 164E

seat back angle, the driver's seat has adjustments for seat cushion height and a firm-to-soft setting for lumbar support. Rear seat passengers won't arrive cramped and crotchety; there is plenty of leg and head room. In all, a most practical package.

Overall, the interior quality is good but there are some inconsistencies. The plastic wood on the center console doesn't match the plastic wood on the dashboard. Also on the dashboard there is a perforated rectangle (for the radio, no doubt) that resembles a cereal-box coupon; we kept looking for the words *cut on dotted line*. That is exactly what will happen when the radio is installed. Better buy the Volvo radio or one that fits that space.

The inertia-reel seat belts are a good feature. Not only are they easy to use but they really work. The last time we

164E manifold has large intake orifice and long runners.

tried these belts was during last year's 142E test and then we found them erratic; one worked some of the time and one didn't work at all. These, however, worked perfectly. It is a one-hand operation to bring them into place; no excuse here for not using them. The belt is rolled in a plastic housing at the base of the center body pillar (no stylish "hardtops" from Sweden!) and threaded up and along the pillar to about shoulder height where it passes through a

ring. One simply reaches over the shoulder and grasps buckle A, pulls it smoothly and slowly across the body and plugs it into receptacle B between the seats. One complaint: the belt tension is too great and it gets tighter and tighter after it is fastened.

Volvo really doesn't need to make the band speedometer accurate; no one seems to know how to read it anyway. Volvo owners tell us they just make an educated guess; some read the pointer of the angled indicator ribbon while others use the area where the ribbon becomes full width. We have had two in a row now that are closest when read at full width and that is where the readings were taken for the data panel.

The dashboard seems bare with only fuel-level and coolant-temperature gauges in addition to the speedometer, odometer and excellent trip odometer with pushbutton reset. Even if the driver never knows what his engine speed or oil pressure is, he will know exactly what the heater is doing because Volvo's heater control knobs are large and lighted in the center to eliminate fumbling at night. Just a glance at the panel tells how much heat is dialed in.

With the vent window open there is some face-level ventilation, but the wind noise makes passengers wish for stale air again. The lower vents provide ventilation below the knees only; it is fitting, however, as they can be operated with the toe of a shoe. We could find no specific instructions for opening the vents but the lever has a rubber pad and looks as if it should be opened and closed with the feet.

Saddled with an automatic transmission, the 164E's performance suffers somewhat. We expect the performance of a 164E with the 4-speed transmission would be considerably better. Even with the automatic it's not underpowered; 0-60 mph in 12.0 seconds and 18.8 seconds at 76 mph in the quarter-mile. It will keep up with traffic in town and cruise quietly past it on the freeway. At low speeds (under 50 mph) the automatic transmission whines softly—probably a normal operating noise. The Pirellis also hum slightly at 65 mph and transmit the freeway expansion-joint bumps rather strongly through the steering wheel. The ZF power steering, by the way, is excellent, as we found in our earlier 164 test.

For states west of the Mississippi the 164E will be the only 164 available; elsewhere in the U.S. both carbureted 164 and the E are on the market—the lingering result of California's threatened 91—octane limit that didn't materialize.

The 164E is a comfortable, medium-size sedan with a price tag of $5050 which includes either the 4-speed manual transmission or the 3-speed automatic, leather seats and trim, radial tires, styled steel wheels and, in general, complete equipment. There are only two options: radio and air conditioning. The 164E is priced just under the BMW Bavaria and significantly below the Mercedes-Benz 250 sedan. The Volvo is quicker than the Mercedes, slower than the BMW, roomier than either, and neither as stylish nor as good-handling as the two German sedans. Overall, it is a fine luxury transportation package that should be durable and won't go out of style tomorrow.

VOLVO 1800ES

*A successful conversion from dated GT
to genuine sportswagon*

WHO'D HAVE THOUGHT the Volvo people would turn the 1800E into a sports wagon? It certainly never occurred to us until the car appeared. The 1800 coupe, now in its 13th year, seemed a candidate for discontinuance but not much else. After all, it was heavy, ugly, cramped, noisy and overpriced. Frank insiders at Volvo didn't equivocate about it: they weren't sure two years ago whether they'd continue the 1800 a while longer, replace it with an up-to-date car, or simply get out of the sporty car business altogether. That business certainly isn't a major part of their activity, although the 1800's design is a good reason it isn't.

But lo and behold, they've had a clever spell, spent a little money on the old coupe and turned it into a sports wagon. And, we must grudgingly admit, the result is amazingly successful. Whoever is responsible for Volvo's station wagon rear sections does nice work indeed, having proved his mettle earlier with the large rear window of the 145 wagon. On the 1800ES he went further with a window that takes up two-thirds of the rear end's height, has hinges and locking handle attached directly to it, and is the entire tailgate. It's futuristic and handsome.

Extension of the roofline straight back put more head-room over what are jokingly called rear seats in the coupe, and in profile view the wagon roof and the well done rear side window give the car such a different profile that many people approached us to ask what the car was, as we toured car-knowing Southern California in it. The whole front end, which was the most acceptable portion of the coupe anyway, is unchanged. The windshield and side glass remain narrow slits, the beltline high and the sides bloated in a most antique way. There's still that corny upsweep in the door that leads to the equally corny chrome-topped fins, but we can't deny that given what they had to work with the designers did a nice job of the transformation.

Sports wagons are by no means a new idea. The MGB, a borderline case because it's so small, has been around since 1966, Chevrolet's first Nomad was based on the Corvette, Reliant builds its Scimitar GTE in England, and there was a handsome Pininfarina one on the Peugeot 504 sports platform last fall. We cannot help being a little parochial, though, and so we note that Volvo's sports wagon, nearly 20 inches longer, roomier, and more wagon-like than the MGB, is the one that's going to put sports wagons on the U.S. map. If the reactions we got from people "on the street" and the fact that Volvo's western U.S. branch is or-

81

LARRY GRIFFIN PHOTOS

ROAD & TRACK
R&T
ROAD TEST

VOLVO 1800ES

dering all its 1800s for 1972 in the wagon body are any indication, it's going to be successful in spite of its dated characteristics and stiff price.

So much for styling critique and market prognostications. What's the car like? Well, as one might have guessed, pretty much like the last 1800 we tested,· the first fuel-injection 1800E two years ago. The engine is still two liters, not 1.8—it mystifies us why Volvo doesn't go ahead and rename the car 2000—but it's one of those that suffered a power cut at the hands of no-lead fuel this year. The rating used to be 124 bhp net with 10.5:1 compression ratio; 10.5 is mighty high and Volvo had to lop off 1.8 points to make the engine run on 91-octane blend, so the power is down to 112 bhp. This shows in two ways. First, the performance is off as expected, with this 1800ES taking 1.2 sec longer to reach 60 mph and 0.7 sec longer to cover the standing quarter-mile than the older car which was 100 lb lighter at test weight. Second, the engine is slightly smoother and quieter than before, a natural consequence of a lower compression ratio. It's still a relatively noisy unit, its pushrod-operated valves clicking classically. It sounds just like the vintage engine it is, but vintage or no it's sturdy and produces good power for its size. Another of our many plugs for electronic fuel injection: it helps this engine to start well from cold, warm up gracefully and run cleanly at all times, its only problem being hard starting when hot.

The 1800's gearbox, inherited from the big Volvo 164, reinforces the engine's sturdy character with its big, heavy shift lever and knob and is a most satisfactory unit. Behind it is the old Laycock-de Normanville overdrive, which engages or disengages hydraulically at the flick of a steering-column stalk and works on 4th gear only. A modern 5-speed box would be better because after slowing down and down-shifting the driver has to do two separate operations or skip a gear on the way back up; but at least the OD provides mechanically relaxed, if not quiet, cruising. A lot of wind noise keeps the car from being quiet at speed.

Seating for the 1800 driver is classic too, with a near-vertical steering wheel, tight dimensions for the shoulders and a "buried" feeling resulting from the relatively high window sills and cowl. The seats are very good, offering not only a backrest angle adjustment but Volvo's unique lumbar-support adjustment. One-operation inertia belts are standard and this was our first test car to have the federally required belt warning: a light on the dash shines until the driver has plugged the belt into its center receptacle between the seats, which is supposed to be lighted but wasn't on the test car. We found no provision for warning the front passenger about his belt.

The rear area encompasses 10.8 cu ft when loaded up to the side window sills; if the jump seats (which are still good only for small kids) are folded down this goes to 14.8 cu ft. And if one is willing to sacrifice rear vision he can stuff nearly 28 cu ft of things into the back end. So the ES is a very capacious 2-seater for long trips, and for local hacking the rear area will be handy too. It always takes a key to get the tailgate open, which is a bother in this sort of use.

Vision outward is far better in the ES than in the coupe because of the large rear glass areas; the super-window at the rear makes it one of the easiest cars imaginable to parallel-park. Up front the slit-like windshield is a problem, however, when the sun visor is needed; it takes up half the windshield's height when it's down! The visor is contoured to conform to the roof shape and its bent-back edge makes it impossible to pivot it down close to the windshield itself—Volvo would do well to adopt a floppy-edged visor.

The ES is a good-riding car over a wide variety of road surfaces and despite having a live rear axle has good suspension travel for large bumps and dips. In fact the car is a little on the soft side and there's a lot of body roll in hard cornering. Combine this with the new and larger Goodyear 70-section tires (same as on the Capri V-6 but in a 15-in. size) and you've got a car that handles well in ordinary-to-brisk driving but gets tippy and squishy at the limit. Thus the benefits of the wider wheels and tires now standard on both the coupe and wagon aren't so significant, at least in comparison with the smaller Michelin tires on our last coupe; the ES has little more cornering power than an average sedan. The steering is precise and reassuring but heavy in parking maneuvers; our test car had a slight steering shimmy at about 50 mph.

The 1800ES is one of those cars that leaves a road-test staff a bit frustrated. Here Volvo has done a nice transformation, produced the first sports wagon big enough to really serve as one that we can buy in America, and done such a nice job with the esthetics that the car is a real head-turner. But they did it on a car that should have been replaced, not reworked. It's a good, solid car but a crude and old-fashioned one; still it's one-of-a-kind and if you must have a sports wagon this is the one you have to choose. We don't think Volvo will have any trouble selling it. 🔘

SCALE: 10" DIVISIONS

PRICE

List price, west coast....... $5032
Price as tested, west coast... $5340
Price as tested includes standard equipment (overdrive, radial tires, leather upholstery, rear window defroster), AM-FM stereo radio ($212), dealer prep ($95)

IMPORTER

Volvo, Inc.
Rockleigh, N.J. 07647

ENGINE

Type............. ohv inline 4
Bore x stroke, mm.. . 89.0 x 80.0
  Equivalent in.... . 3.50 x 3.15
Displacement, cc/cu in.. 1986/121
Compression ratio.......... 8.7:1
Bhp @ rpm, net. . 112 @ 6000
  Equivalent mph... 124
Torque @ rpm, lb-ft... 115 @ 3500
  Equivalent mph....... 74
Fuel injection..... Bosch electronic
Fuel requirement... regular, 91-oct
Emissions, gram/mile:
  Hydrocarbons................ 1.5
  Carbon Monoxide........... .28
  Nitrogen Oxides........... 2.3

DRIVE TRAIN

Transmission: 4-speed manual plus overdrive
Gear ratios: OD (0.797).... 3.43:1
  4th (1.00)................ 4.30:1
  3rd (1.36)................ 5.85:1
  2nd (1.99)................ 8.56:1
  1st (3.13)............... 13.45:1
Final drive ratio.......... 4.30:1

CHASSIS & BODY

Layout..... front engine/rear drive
Body/frame............ unit/steel
Brake system: 10.6-in. disc front, 11.6-in. disc rear; vacuum assisted
  Swept area, sq in......... 400
Wheels..... styled steel, 15 x 5½ J
Tires............. Goodyear G800
  185/70 HR-15
Steering type........ cam & roller
  Overall ratio............ 15.5:1
  Turns, lock-to-lock........ 3.2
  Turning circle, ft.......... 31.5
Front suspension: unequal-length A-arms, coil springs, tube shocks, anti-roll bar

Rear suspension: live axle on upper & lower trailing arms with Panhard rod; coil springs, tube shocks

ACCOMMODATION

Seating capacity, persons.... 2+2
Seat width, front/rear.2 x 19.0/41.0
Head room, front/rear.. 35.5/35.5
Seat back adjustment, degrees.. 40

INSTRUMENTATION

Instruments: 120-mph speedometer, 7000-rpm tach, 99,999 odometer, 999.9 trip odo, oil press, oil temp, coolant temp, fuel level, clock
Warning lights: ammeter, high beam, directionals, hazard flasher, handbrake, overdrive, seatbelt

MAINTENANCE

Service intervals, mi:
  Oil change................ 6000
  Filter change............. 6000
  Chassis lube............. 6000
  Minor tuneup............. 6000
  Major tuneup........... 12,000
Warranty, mo/mi..... 6/unlimited

GENERAL

Curb weight, lb............. 2570
Test weight................ 2935
Weight distribution (with driver), front/rear, %... 50/50
Wheelbase, in.............. 96.5
Track, front/rear........ 51.6/51.6
Length................... 172.6
Width.................... 66.9
Height.................... 50.4
Ground clearance........... 6.1
Overhang, front/rear... 30.2/45.9
Usable trunk space, cu ft..... 10.8
Fuel capacity, U.S. gal........ 11.9

CALCULATED DATA

Lb/bhp (test weight)........ 26.2
Mph/1000 rpm (o'drive)..... 21.4
Engine revs/mi (60 mph o'drive)......... 2800
Piston travel, ft/mi........ 1470
R&T steering index........ 1.02
Brake swept area, sq in/ton.. 272

RELIABILITY

From R&T Owner Surveys the average number of trouble areas for all models surveyed is 11. As owners of earlier-model Volvos reported 10 trouble areas, we expect the reliability of the Volvo 1800ES to be average.

ROAD TEST RESULTS

ACCELERATION

Time to distance, sec:
  0-100 ft.................... 4.0
  0-500 ft.................... 9.9
  0-1320 ft (¼ mi)....... 18.2
Speed at end of ¼-mi, mph.... 74
Time to speed, sec:
  0-30 mph................. 3.7
  0-40 mph................. 5.6
  0-50 mph................. 8.0
  0-60 mph................ 11.3
  0-70 mph................ 15.5
  0-80 mph................ 21.4
  0-90 mph................ 30.1

SPEEDS IN GEARS

O'drive (5600 rpm).......... 116
4th (6500)................. 108
3rd (6500)................. 80
2nd (6500)................. 56
1st (6500) 36

SPEEDOMETER ERROR

30 mph indicated is actually.. 28.5
50 mph.................... 47.5
60 mph.................... 57.0
70 mph.................... 66.5
80 mph.................... 75.0
Odometer, 10.0 mi......... 9.9

BRAKES

Minimum stopping distances, ft:
  From 60 mph.. 178
  From 80 mph.. 299
Control in panic stop........ good
Pedal effort for 0.5g stop, lb.... 20
Fade: percent increase in pedal effort to maintain 0.5g deceleration in 6 stops from 60 mph....... 50
Parking: hold 30% grade?... yes
Overall brake rating..... . good

HANDLING

Speed on 100-ft radius, mph. 31.5
Lateral acceleration, g....... 0.660

FUEL ECONOMY

Normal driving, mpg........ 22.5
Cruising range, mi (1-gal res.). 245

INTERIOR NOISE

All noise readings in dbA:
Idle in neutral................ 57
Maximum, 1st gear..... . .. 84
Constant 30 mph....... 65
  50 mph.................... 75
  70 mph.................... 77
  90 mph.................... 82

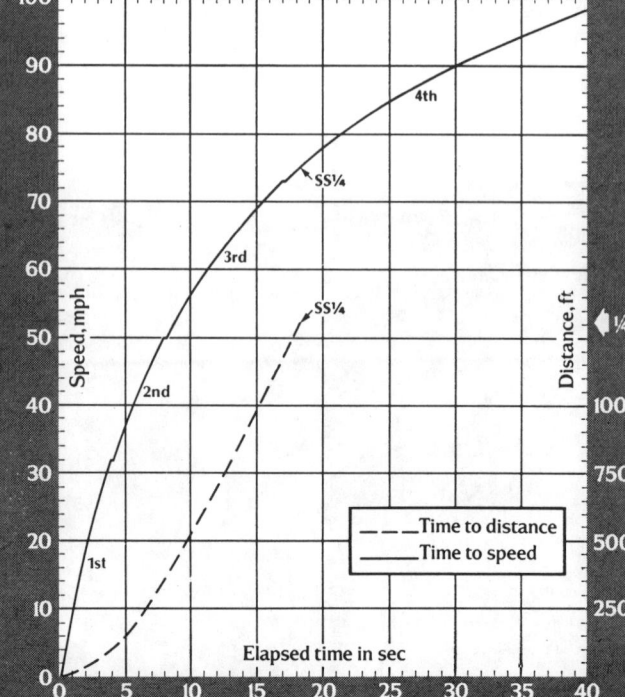

ACCELERATION

Elapsed time in sec

VOLVO BUILDS A SAFETY CAR

A maker with a long history of interest in safety ventures into crashworthiness

BY RON WAKEFIELD

VOLVO, A BUILDER of practical, no-nonsense cars, also has a reputation for building "safe" cars. Volvo pioneered the 3-point seatbelt before 1960, had a collapsing steering column long before it became a legal requirement, devised a particularly effective circuitry for fail-safe braking hydraulics, etc, etc. Fact is, if U.S. manufacturers had taken the attitude Volvo traditionally has about safety we probably wouldn't be suffering from over-legislation as we are today.

Be that as it may, Volvo's reputation made it imperative for the Swedish company to at least make an effort to meet the U.S. government's ill-considered crash safety requirements as laid down in the ESV (Experimental Safety Vehicle) project. The firm had already begun its Safety Car project in 1969, but when the ESV program was announced in 1970 it influenced the direction of Volvo's program.

Today the requirements for Volvo's project and the ESV are quite similar in several important respects, including the 50-mph barrier crash test, 15-mph side impact test, 50-mph pole crash, front bumper that protects the body from damage in a 10-mph barrier crash, etc. Others are different: Volvo has different ideas about what constitutes good handling, lighting and instrumentation-control layout. The target weight for Volvo's car was 3000 lb (1360 kg) and it, like the ESVs, had to meet 1974 U.S. exhaust emission regulations. At its present stage of development the car, of which there are two running prototypes, weighs about 3200 lb, meaning that it is much closer to its target than are the U.S. Fairchild, AMF and GM ESVs to their 4000 lb. Whether it meets its crashworthiness targets is not yet known.

In general design the Volvo is much like the NHTSA's ESV. Its body-chassis structure is an extremely robust unit

In this prototype an airbag lives in the dash's right end, ready to blow up in the passenger's face. Huge seatback bolsters, right, with pop-up head restraints protect front and rear . . .

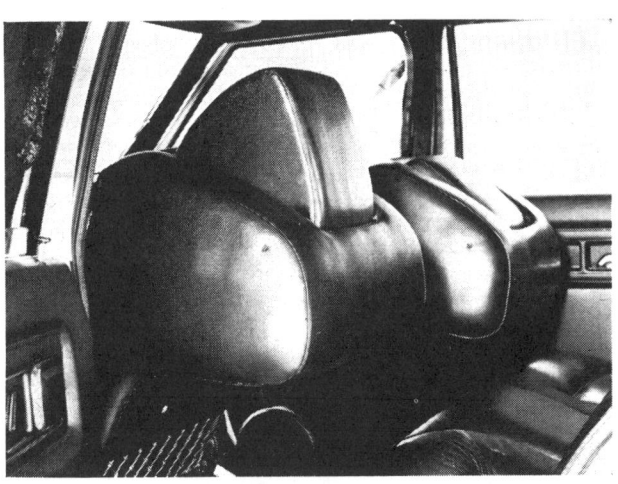

. . . passengers but those in rear probably wouldn't like them.

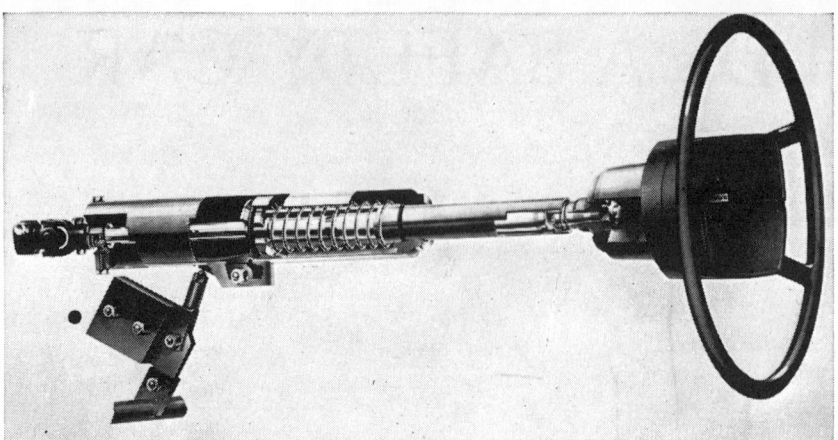

Impact-actuated steering column has preloaded spring that is freed to pull the steering wheel forward 6 in. Steering is retained even in this event.

One-way "passive" 3-point belts move into place automatically but must be put away when occupants exit.

VOLVO SAFETY CAR

of steel members with its ends designed specifically for energy-absorbing collapse. Doors carry tubing that interlocks with body sides when they are closed. It's a 4-door, 4/5-passenger sedan on a 106.3-in. wheelbase (same as today's Volvo 164) and much wider front and rear tracks than any of today's Volvos. It's 206 in. long, 71.7 in. wide and 55.8 in. high, respectively about 20 in. longer, 3 in. wider and an inch lower than the 164. The length increase is all in the bumpers, as the photos show and as is typical of safety-car design. Suspension is similar to current Volvo practice with unequal-arm independent at front, a live axle on links at the rear; springing is by coils all around and there is automatic leveling on one of the test cars by compressed air. Brakes are disc and they're prevented from locking up in panic braking by an electronic anti-skid system. The engine is a 4-cyl B20 unit with electronic fuel injection, an air pump, exhaust gas recirculation and a catalytic reactor, driving through a normal Volvo 4-speed manual transmission. The car's performance must be something to behold, with the 4-cyl engine giving about 95 bhp to drive the 3200 lb, but use of the 6-cyl 164 engine (which would develop around 125 bhp with the emission mods) was probably ruled out by the need for "crush" space up front.

Volvo's research into the tolerance of the human body for deceleration loading indicates that a man must have a minimum of 4 ft 11 in. in which to decelerate from 50 mph if he is to survive a crash at that speed. That's an interesting figure; the U.S. government safety agency (NHTSA) chief said 3 ft, but all the ESVs are obviously designed for much more and so is the Volvo. Volvo does it by providing a deformation zone of 3 ft 7 in. at the front and letting the passenger move forward with controlled motion about 16 in. before he is fully restrained by the restraint system, whatever that be. When you consider that today's largest cars have about 2-2½ ft of available crush space up front, it becomes clear that any 50-mph crash car just has to be very long and/or wasteful of space.

As for restraint systems, one of the Volvo prototypes has a "semi-passive" belt system by which the 3-point belts are automatically pulled across the passengers after the doors are closed (see photo); the passengers must release the belts before getting out, which saves some mechanical complication. The belts are a bit wider than today's standard ones. Volvo's safety director Rolf Mellde considers belts to be the best safety feature in today's cars but does not consider their possibilities to have been fully developed. R&T agrees completely; they are the only restraint system whose value

is proven and we believe there is considerable scope for improvement on them.

The other prototype has airbags for all occupants—even one on the rear parcel shelf which deploys in a rear-end collision for head restraint there. Volvo has little to say about airbags, but of course with the NHTSA's intriguing prejudice for them a carmaker has to at least try them. Padded, pop-up head restraints are also being tried as an alternative to the airbag ones. A further idea in crash-deployed safety devices is a steering column that is forced about 6 in. forward upon frontal impact by a spring—an interesting extension of the collapsible column principle and a way to "buy" extra deceleration distance for the driver.

Volvo's styling job on the safety car is the most palatable one we've seen so far. The 10-mph bumpers—the front one must take a 10-mph barrier crash, the rear 10 mph into another car—are inevitably an atrocity, looking worse than a bent front or rear end and surely making parallel parking next to impossible, but the body lines on this car are clean and crisp. The bumpers are on telescopic mountings.

The Volvo safety car is not an ESV—that is, it was not developed under contract to the American government. Volvo is, however, under agreement to the U.S. on the anti-skid brake system, developing it and studying the maneuvering capacities of braked, but not skidding, vehicles for possible feed-in to the NHTSA. What the Volvo safety-car project, as well as the American ESV project, should accomplish is to show what level of safety can be built—and cannot be built—into a real car for real people to buy and drive. Volvo's design is not that, but at least it is the most compact and lightest experimental safety car to be built so far. ◉

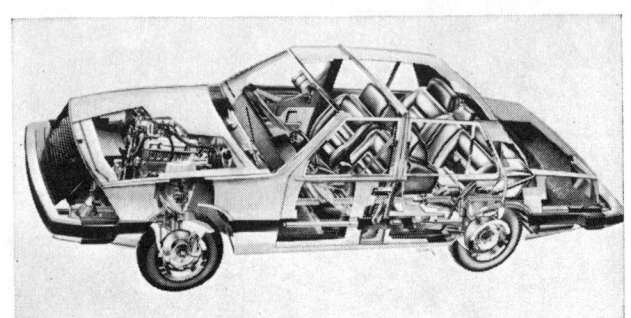

Bumpers for 10-mph barrier crash (front) and 10-mph car-to-car crash (rear) together account for about 11 in. of car's 206-in. length; 4-cylinder engine also helps get "crush space."

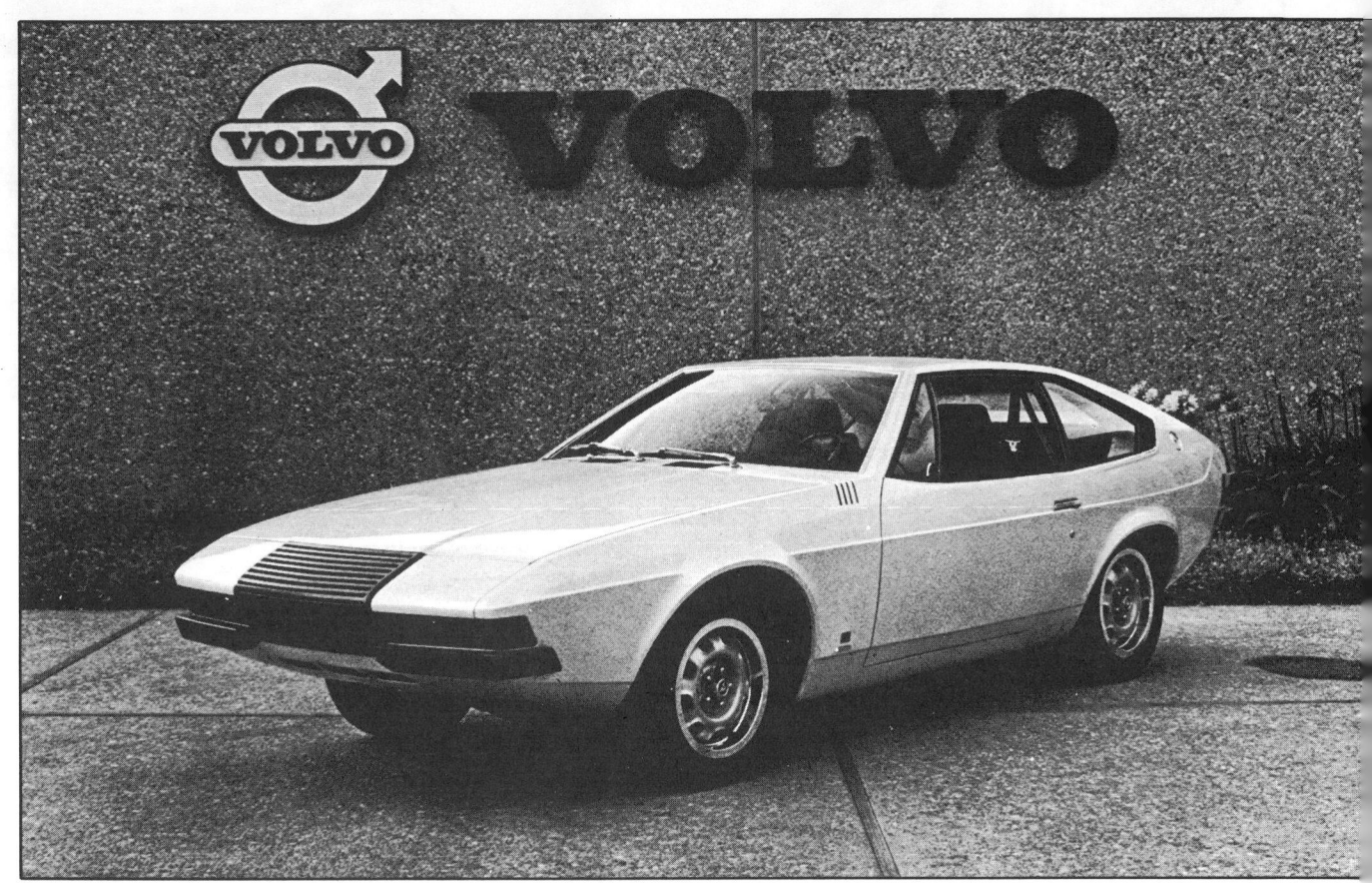

New from Sweden & Italy:
COGGIOLA VOLVO 1800ESC

This practical but exciting design by a little-known Italian designer is a logical way for Volvo's sports model to go

BY RON WAKEFIELD

PHOTOS BY JOE RUSZ

W HEN VOLVO brought out the 1800ES sportswagon version of their venerable 1800 late last year we relaxed back into our editorial chairs, sighed and resigned ourselves to another several years of the sturdy but old-fashioned series. The coupe was barely stylish when it was introduced in 1960, and though Volvo had mechanically updated it many times with more displacement here, more disc brakes there, and electronic fuel injection, there remained the cramped, poorly laid-out body. True, the wagon version was a big improvement but it didn't quite succeed in converting the sow's ear.

A few show cars on the 1800 chassis—still, curiously, called 1800 by Volvo even though the engine was increased to two liters some time ago—have appeared from time to time. Zagato showed a bland-looking one at Geneva last year, for instance. But this year's Paris show brought a new one—this time one commissioned and paid for by Volvo itself, which can only mean that something's up in Göteborg. Are those nice people at Volvo finally thinking of a fresh new body for this fine old chassis? Of course we hope so, and the new show car shows every sign of being a serious study of the possibilities. When

it came to Los Angeles to be shown at L.A.'s finest motor show, Auto Expo, we took a close look at it and liked what we saw.

The builder, Sergio Coggiola, is a former Ghia designer now on his own. He's actually been running his own design firm since 1966 and has done projects for car companies in Europe and Japan, but this is the first complete car we've seen from him.

First let us emphasize that the Coggiola 1800ESC, as it is called, is no far-out show car. No, it is a practical car, an exercise not in gimmickry but in getting more useful space out of essentially the same package size as today's 1800E *and* giving it a modern look. Mind you, the name ESC implies a relation to the 1800ES—the wagon—and that's what Volvo people like to refer to when they're comparing it to the present model. But the ESC is clearly a coupe, not a sportswagon, and for this reason we prefer to compare it to the production coupe.

The Coggiola body resides on the 1800 chassis—wheelbase, track, suspension pieces, engine, drivetrain, floor platform and all. But all the sheet metal and the body layout are as new as using the existing chassis would allow. Coggiola has kept exactly the same overall height as the production car—which

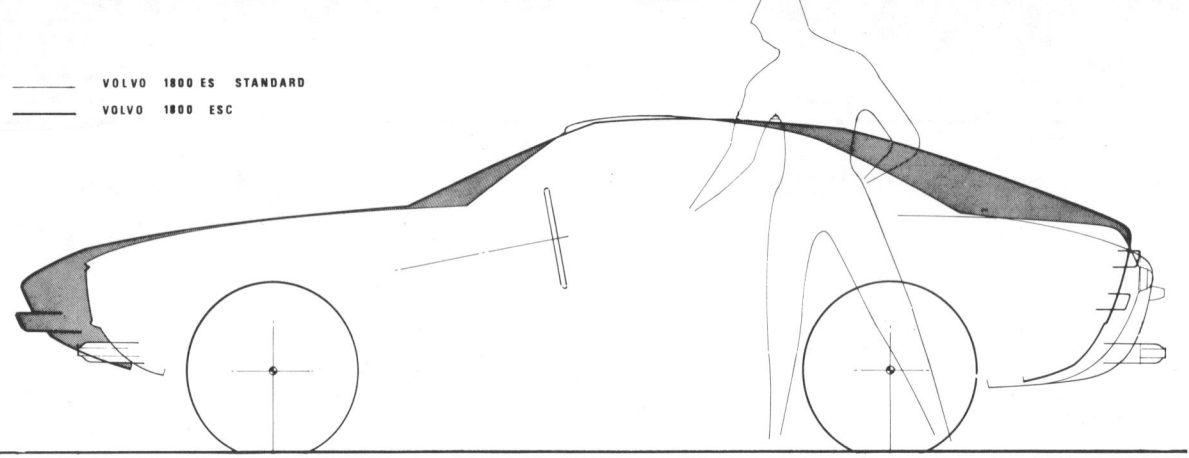

VOLVO 1800 ES STANDARD
VOLVO 1800 ESC

is plenty low—but has gone five inches longer while reversing the balance of front to rear overhang. Thus the nose is long and the tail short, following both current styling trends and the direction future safety requirements dictate. The roof drops off less abruptly than that of the current coupe, thus carrying full headroom a bit farther back over the passenger compartment, but the rear seating is still occasional in the fullest sense of the word.

The lines are handsome if not especially original—there's a strong flavor of recent Bertone and Giugiaro themes and the car is even reminiscent of that earlier Zagato effort. A strange, but functional, gray-painted grille wraps up into the long, sloping hood and is flanked by matching semi-bumpers (hardly practical in view of U.S. bumper regulations coming up) and hiding headlights that are raised by hydraulic units responding to a pedal inside the car. The light units come straight up and Coggiola perhaps declined a bit of corny cuteness by not finishing off their exposed edges airfoil-style.

Body sides are very much in the Giugiaro idiom of a year or so ago, with the upper body structure blended smoothly into the lower, the window ledge dropping well down into the lower body, and a crease originating across the front carrying itself all the way to the rear with only an interruption at the front wheel arch, finally kicking up to repeat the quarter-window motif. The cowl is high, almost as if to remind us that this is an 1800 Volvo; the pushrod, vertically mounted engine is fairly high but it seems that only Coggiola's determination to make a virtually straight slope from nose to windshield has necessitated its being as high as it is.

The standard 185/70-15 tires aren't overly emphasized in the design and are mounted on regular 1800 wheels. Repeating the quarter-window lines are ventilation flow-through outlets behind the windows, and at the rear the bumpers are doubled up but still impractically close to the bodywork. Nearly everything on the tail is clean too, but Coggiola succumbed to

temptation here with what we might call simulated louvers under the rear window. These are part of a conventional decklid that ends beneath the window rather than including it.

Inside, the low window ledge is much appreciated even though the high cowl (here again) reminds one of the old 1800. All the instruments are straight from the production car (and are nice), set into a clean, contemporary panel with a full-width brushed aluminum band that's matched by the central console. The 1800's large, traditionally vertical steering wheel obscures one or two of the gauges, depending upon driver position. Switches for non-driving functions are laid out within easy reach of the driver above radio, heater and what would be air-conditioning controls if the car actually had it (the innards are missing), and the steering column presents three stalks for more critical driving functions. Surprisingly, there are no face-level air vents.

The two front seats are roomier in feel and in fact than the production car's—extra headroom and, even more so, shoulder room are noticeable. The +2 seats behind, though, almost might

DIMENSIONS

	Production 1800E	Coggiola 1800ESC
All dimensions in inches:		
Wheelbase	96.5	96.5
Track, front/rear	51.6/51.6	51.6/51.6
Length	171.3	176.3
Overhang, front/rear	30.2/44.6	40.8/39.0
Width	66.9	68.2
Height	50.6	50.6

COGGIOLA VOLVO 1800ESC

as well be left out, just as in the current model—there's less headroom there than in the current sportswagon though more than in the coupe. Seat upholstery is much overdone and the armrests on the doors carry out the same silly theme, but door panels themselves and everything else are simple and elegant.

The gearshift is far away, as are the pedals, and the steering wheel too close—more evidence of the old 1800 mechanical arrangement and surely something that would be changed if the car were put into production. In fact, every mechanical control is right where it is in today's car, including the odd outboard handbrake lever.

The rear seats are flanked by wheelwells, and forward of these are large cubby boxes that could be useful storage. Inertia-reel seatbelts feed out of the interior panels nearby. At the aft end is a carpeted floor under which the spare tire lives, not encompassing a lot of luggage space, but things can be set upon or stuffed into the occasional seats. The live rear axle doesn't cost a lot of room here; it requires a full-width rise across the front of the luggage area but this is only a few inches lengthwise.

The pictures tell their own esthetic story. We like the Coggiola Volvo very much and consider it a perfectly logical basis for a new production 1800—let's go ahead and call it a 2000—coupe. Sure, it needs a few changes: driver position is one, the bumpers another. And since new tooling would inevitably drive the price even higher than the present $5000, Volvo might well question using a pushrod inline 4-cyl engine and consider a compact V-6 to reduce height and give more refined running. This could use the four's pistons, rods, valvegear and so forth, give an attractive 3-liter capacity and do away with the need for overdrive.

How about it, Volvo?

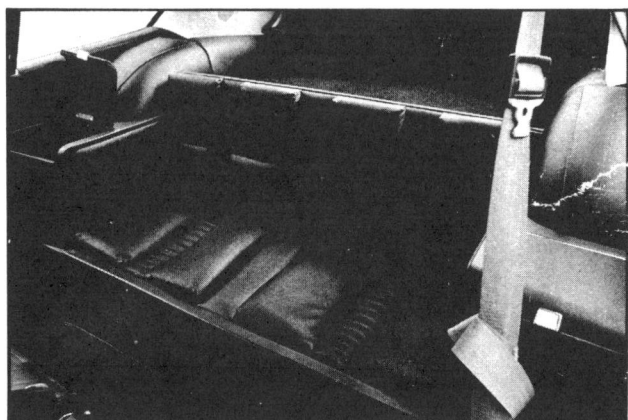

VOLVO 142GL

*Volvo's latest 4-cylinder
is an improved car*

VOLVO'S VENERABLE SERIES of 4-cylinder cars—142 2-door sedan, 144 4-door sedan and 145 station wagon—has been around since 1966. Over the years these tall, boxy cars have gained a reputation for being solid, safe and spacious if a little stodgy. Our last experience with a 140-series Volvo—in a comparison of the Audi 100LS, Peugeot 504, Saab 99E and Volvo 144E in November 1972—was not a particularly pleasant one. We rated the Volvo last for several reasons, not the least of which were its old-fashioned instrumentation, controls and ventilation and its great reluctance to get moving from rest. These problems were taken care of in 1973: the instrument panel was completely redesigned and a higher numerical 1st-gear ratio was fitted so that the car no longer fell flat on its face when starting from a stoplight.

For 1974 Volvo has further improved the 140 series and added two new models to the U.S. line. Changes are mostly of a comfort and safety nature and include reinforced one-piece door-window frames, a fuel tank relocated forward for added protection in a rear impact, enormous (and ugly) shock-absorber bumpers front and rear, ventless side windows for less wind noise, and power-assisted steering as standard equipment on all models with automatic transmission. The additions to the 140 series are the 142GL (subject of this road test) and 144GL, cars that have been available in Europe for some time. In a sense these Grand Luxe models are hybrids: 140s with some of the interior appointments of the 6-cylinder 164. To an already impressive list of standard equipment Volvo has added a steel sunroof, leather-faced upholstery, heated driver's seat, tachometer, styled steel wheels and steel-belted radials (fabric-belted radials are standard on other 140 models) and a standard overdrive behind the 4-speed transmission.

The Volvo interior, always noted for its roominess and comfortable multi-adjustment seats, is an even nicer place now. Last year controls were better arranged, that dreadful ribbon speedometer that nobody knew how to read was replaced with a more conventional type and the radio was positioned within reach of the driver. This year there are ventless windows to go with the greatly improved vent system introduced in 1973 and wind noise, having been bad in 140s, is reduced. Standard for all models is a warning light which comes on if a low beam, brake light or taillight burns out—a thoughtful safety touch.

There's little difference in the way the 142GL rides and handles: it's typically Volvo with a lot of roll in corners (the high seating enhances the feeling), final oversteer, better-than-average wheel travel and a firm ride. The steering, too, feels little different despite a change in 1973 to reduce effort (our car didn't have power steering). Turning effort is still fairly high and the steering is vague, lacking feel particularly when cornering. We place most of the blame on the relatively soft front springs; two Volvos we've driven recently equipped with handling packages had much more responsive and precise steering. Stopping distances are also about normal, but there was considerable rear locking during our stops from 80 mph and more fade than we expect to find in a 4-wheel disc system.

The one mechanical change of note concerns the engine. There's nothing new in the basic design—it's the same pushrod 2-liter, first introduced in the 1800E back in 1970 and in the 140 series a year later. But Volvo, like Porsche, has adopted Bosch's new mechanically timed K-Jetronic injection system in place of the also-Bosch electronic system used previously. Though the throttle/injection linkage is a bit stiff and notchy, throttle response is good. The engine starts on demand hot or cold and runs without stalling or stumbling, and there's no lean surge to mar its light-throttle performance. There's good acceleration through the gears—the numerically higher 1st gear really helps—and Volvo has done a decent job of maintaining performance in the face of weight increases and more restrictive emission regulations. In fact, although this year's engine is rated at 3 bhp lower than the last 140 we tested it's appreciably faster: almost 2 sec quicker to 60 mph and 1.3 sec faster in the quarter-mile. And this despite an increase in curb weight of 130 lb. But a nasty vibration period at 2500 rpm that we don't recall in past B20 engines set off a racket in the dash of the test car.

Overdrive is a worthwhile addition, making cruising quite silent and giving fuel economy at 70 mph that should embarrass 55-mph do-gooders. We ran this car through our normal fuel-economy course twice, first using 70 mph for our freeway sections (it was legal then) and a second time keeping to a 55-mph limit. In the second instance overdrive cruising fell victim to legislative overkill, yielding only 18 mpg vs the 20 mpg we obtained when using 70. This is no surprise, said the Volvo people. Using overdrive at 55 mph would be all right if the road were perfectly level, but even a slight incline requires a wide throttle opening to maintain 55 mph and fuel economy goes sour. This leaves the GL owner with a dilemma: drive "illegally" at 65-70 mph in overdrive, suffer with lower fuel economy at 55 mph in overdrive or leave the transmission in 4th, drive at 55 mph and put up with additional engine noise. Once again, blanket laws triumph over logic in particular situations.

Volvo prices, like everything else it seems these days, have been caught in an upward spiral that finds the 142GL costing more than a 164 did only a few years ago. Many buyers, however, will find the GL an attractive alternative to the 164, which offers only an extra bit of luxury, performance and refinement for its $1500-higher price tag. Overall, the 142GL is a comfortable medium-size sedan with an extensive list of standard equipment and impressive fuel economy. If its bulk and height make it a little cumbersome to drive they also make for generous interior dimensions and a huge trunk. It's an utterly practical luxury car and one that should be durable and in style for many years to come.

PRICE

List price, west coast $5465
Price as tested, west coast .. $6065

ENGINE & DRIVE TRAIN

Type	ohv inline 4
Bore x stroke, mm	88.9 x 80.0
Displacement, cc/cu in	1990/121
Compression ratio	8.7:1
Bhp @ rpm, net	109 @ 6000
Torque @ rpm, lb-ft	115 @ 3500
Fuel requirement	regular, 91-oct
Transmission	4-sp manual with OD
Gear ratios: OD (0.80)	3.43:1
4th (1.00)	4.30:1
3rd (1.36)	5.85:1
2nd (1.99)	8.56:1
1st (3.41)	14.66:1
Final drive ratio	4.30:1

CHASSIS & BODY

Body/frame	unit steel
Brake system	10.7-in. disc front, 11.6-in. disc rear; vacuum assist
Wheels	styled steel, 15 x 5J
Tires	Michelin ZX, 165SR-15
Steering type	cam & roller
Turns, lock-to-lock	3.8
Suspension, front/rear: A-arms, coil springs, tube shocks, anti-roll bar/live axle on trailing arms & Panhard rod, coil springs, tube shocks	

GENERAL

Curb weight	2870
Weight distribution (with driver), front/rear, %	52/48
Wheelbase, in.	103.0
Track, front/rear	53.1/53.1
Length	188.0
Width	67.1
Height	56.5
Fuel capacity, U.S. gal.	15.8

CALCULATED DATA

Lb/bhp (test weight)	27.6
Mph/1000 rpm (OD)	21.1
Engine revs/mi (60 mph)	2850
R&T Steering index	1.20
Brake swept area, sq in./ton	266

ROAD TEST RESULTS

ACCELERATION

Time to distance, sec:
0-100 ft	4.0
0-500 ft	10.2
0-1320 ft (¼ mi)	18.7
Speed at end of ¼ mi, mph	73.0

Time to speed, sec:
0-30 mph	4.1
0-50 mph	9.0
0-60 mph	12.7
0-80 mph	24.5

SPEEDS IN GEARS

OD (4800 rpm)	102
4th (5800)	102
3rd (6500)	79
2nd (6500)	54
1st (6500)	31

FUEL ECONOMY

Normal driving, mpg 20.0

BRAKES

Minimum stopping distances, ft:
From 60 mph	165
From 80 mph	297
Control in panic stop	fair
Pedal effort for 0.5g stop, lb	20

Fade: percent increase in pedal effort to maintain 0.5 deceleration in 6 stops from 60 mph 50
Overall brake rating .... good

HANDLING

Speed on 100-ft radius, mph	31.4
Lateral acceleration, g	0.658
Speed thru 700-ft slalom, mph	49.4

INTERIOR NOISE

All noise readings in dBA:
Constant 30 mph	64
50 mph	68
70 mph	75

SPEEDOMETER ERROR

30 mph indicated is actually	27.0
60 mph	58.0
70 mph	67.0

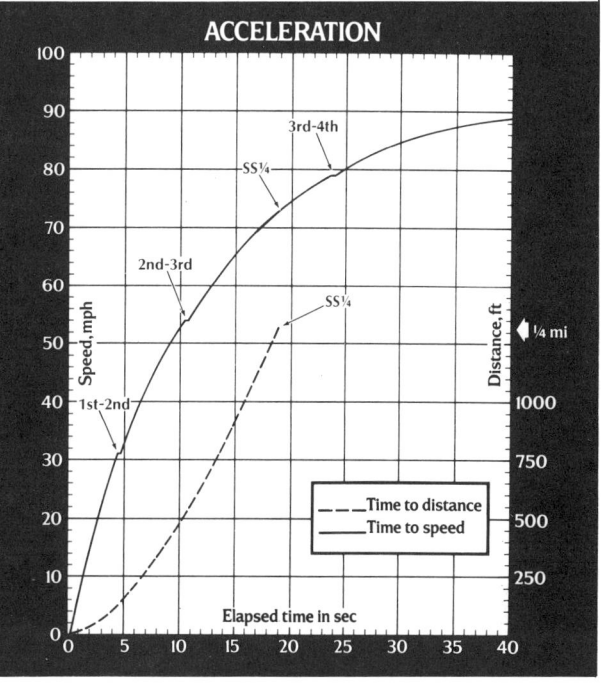

ACCELERATION

THE VOLVO SEDANS

*Good things from Volvo Competition Service
& Import Parts Distributing for 140s & 164s*

BY JOHN DINKEL, Engineering Editor

Mention Volvo and most people think of nothing but sturdy, practical transportation. But Volvos can also be fun to drive. All it takes is the right combination of performance accessories—as we discovered after a week of driving and testing a pair of modified Volvos. We tried a highly modified 1971 142E from Import Parts Distributing Company (2762 NE Broadway, Portland, Oregon 97232; 503 287-1179) and a 1974 142 equipped with a few of the numerous heavy-duty, off-road and competition parts now offered by Volvo through its Competition Service Department (1955 190th St, Torrance, Calif., attention: Wayne Baldwin; 213 770-1550).

Import Parts Distributing is probably an unfamiliar name to all but hardcore Volvo competitors. A small group of dedicated enthusiasts, IPD has been involved with Volvos since the early 1960s, starting with a 544 at the drags and progressing to a fuel-injected 1800 coupe for road racing. In 1965 IPD sponsored a 122S that won the B-Sedan title in SCCA's Northern Pacific Division, raced another B sedan in 1969 and ran an E-Production 1800 in 1970. Currently in the works for sedan racing is a 142, a car IPD thinks will make the most competitive Volvo race car to date.

Because no serious factory-supported or even independent Volvo racing effort existed in this country prior to IPD's involvement, they had to pretty much start from scratch in designing and building competition equipment. But a wealth of experience and information IPD has gained the hard way—on the race track—has resulted in the development of a complete line of parts and accessories for improved performance. The IPD catalog lists special engine components such as valves,

cylinder heads, forged pistons, oil-pan baffle, cams and valve springs; and chassis components such as front anti-roll bars and adjustable rear bars, special coil springs for race and rally, and polyurethane suspension bushings for fine-tuning the chassis. To get that extra power to the ground IPD sells limited-slip differentials, wide wheels, special gear ratios and overdrive units plus heavy-duty clutches and light aluminum flywheels. If you're a competition-minded Volvo owner get the catalog; it's a dollar well spent.

According to IPD the stock Volvo drivetrain and suspension are so strong that generally no reinforcing or special materials are necessary, making a Volvo one of the least expensive cars to prepare for racing. That's why many parts are absent from the IPD catalog: heavy-duty connecting rods, rocker arms and crankshafts to name just three. However, if you need specific information on some component or modification not listed in the catalog, IPD suggests you phone for technical advice: until 5:30 PM PDT weekdays and until noon on Saturday for you weekend tinkerers.

Volvo's Competition Service Department was announced about a year ago but is just now reaching the stage where dealers nationwide are aware of the program and have received a copy of Volvo's very complete Competition Service Catalog. The first part of the book is devoted to FIA homologation papers on the various Volvo models so that you can determine exactly which modifications and parts are legal. Next comes a section on competition parts, including engine tuning kits, exhaust systems, limited-slip differentials, suspension components, skid plates for off-road excursions, racing bucket seats, instruments

and gauges, fiberglass body panels and just about every nut and bolt you could possibly need to rebuild your Volvo from the ground up. The last section contains detailed performance-tuning information for the current B20 engine: specifications, running adjustments and several photos to explain procedures and point out exactly which nuts and screws to adjust and by how much. Although not every dealer has an inventory of competition parts, each does have a copy of the Competition Service Catalog and can quote prices and order parts from the Competition Service Department in California. Rather than attempt to list all the parts available, we suggest any interested Volvo owner visit his local dealer and ask to see the competition parts catalog.

Two Examples

To demonstrate how well a Volvo can go, ride and handle Dick Gordon and Rich Meyer of IPD journeyed down from Oregon in Gordon's 142E sedan. At the same time we arranged to borrow from Volvo Western Distributing a stock 1974 142GL (a road test of the car appeared in our April issue) and a mildly tuned 142 equipped with a few of the pieces listed in the Competition Service Catalog. The parts fitted to each of the tuned cars and the cost is found in the Performance Equipment table. The test-data table follows with a comparison of the three cars for acceleration, braking and handling. Keep in mind that the engines in both of the tuned cars bear no relation to ones that would meet anyone's emission laws, so we really can't recommend these modifications for anything except competition and other off-highway work.

Performance modifications to Volvo's 142 consisted of a rally camshaft, headers and a 4.56:1 final drive. The camshaft proved a mixed blessing, improving top-end performance but sacrificing low-speed tractability and throttle response. If the stock compression ratio is to be retained (as it was in this instance) one of the milder Volvo cams with less overlap would probably be a better choice. We had mixed reactions to the final drive ratio as well: it's an advantage for drag-racing starts but is too high numerically for comfortable freeway cruising even at 55 mph. For highway usefulness it should be combined with overdrive.

For improved handling, wider alloy wheels, 185-15 radials (Pirellis, in place of the stock 165-15 Michelin ZXs), stiffer front springs and a larger front anti-roll bar were installed. Roll, notoriously bad in Volvo sedans, was considerably reduced by the springs and bar; and for some reason the steering became noticeably lighter as well as more precise. The stiffer front springs reduced Volvo's characteristic oversteer too, but also introduced some pitching; we suggest purchasing matching rear springs to reduce this tendency. This would of course restore the oversteer—not necessarily a bad thing.

Stopping distances were an enigma. Panic stops from 80 mph were no better than stock and from 60 mph took 15 feet longer. The stiffer front springs result in much less dive and better control and the wider wheels and tires should help too, so perhaps the Pirellis despite their larger size didn't deliver the stopping traction of the Michelins. There's no questioning the modified Volvo's advantage in handling, however: both the lateral acceleration and slalom speeds are great improvements

93

Dick Gordon (right) and Rich Meyer of IPD, performing some on-the-track modifications to their Volvo at Orange County Raceway.

Three levels of behavior in the R&T slalom test: the stock 142GL...

the Volvo Performance Service 142, looking clearly less ruffled...

and the IPD car, looking still better. Speeds are given in the table.

For dark rally roads: these Bosch lamps (55-watt fog inboard, 100-watt driving outboard) are available from Volvo Competition Service.

over the 142GL's capabilities.

The highly modified IPD Volvo was, as expected, considerably quicker than either of the other two cars. For one thing, valvetrain modifications allowed the engine to be revved to 7000 rpm. We shifted at 6800 to be on the safe side, but there's usable power all the way to the redline and yet the engine is very tractable at low speeds as well. An interesting modification, and one unique to IPD as far as we know, is an adjustable computer box for the Bosch electronic fuel injection. For $75 IPD will modify a standard "black box" so that the mixture can be adjusted by the driver from 10% lean to 30% rich in increments of 5%. This allows the fuel-air ratio to be set—while driving—for maximum economy on the street or for maximum power when racing. Twisting the dial as we drove, we could feel the difference: the engine would begin to surge if we went too lean and would accelerate more strongly with a rich setting.

Though the increased power makes the IPD Volvo a much more responsive car to drive, it's the suspension modifications that work the greatest transformation on the Volvo's character. To appreciate what IPD has achieved you have to be an enthusiast: the ride with the shorter, stiffer IPD coil springs, 6¾-in.-wide American Racing Equipment 200-S aluminum wheels and Semperit 185/70-15 radials is definitely on the firm side. It's not uncomfortably harsh, however, even over rough roads as we'd normally expect with suspension that's been both lowered and stiffened. IPD credits the Bilstein gas-pressure shock absorbers for the Volvo's fine ride and R&T concurs. Good things have happened to the steering and brakes also. Normally vague and rather unresponsive, the Volvo's steering with the IPD mods is positive and direct with no lost motion, so the car was easy to drive fast and smoothly through our slalom course. Improvements in braking result from a combination of factors: the stiffer springs, which reduce front-end dive, the wider Semperit tires and the Repco brake pads.

Gordon's Volvo didn't disappoint us on the skidpad either: a lateral acceleration of 0.785 puts the car among a select group of expensive sports cars. And its transient behavior is even more impressive. The car has excellent balance with flat, neutral response changing to mild, controllable oversteer as the limit is reached. Most important, this car was more at home on twisty roads than any other Volvo we have driven.

Normally, in an Improved Performance article we list all the manufacturers we know of who have equipment available for the make. In this instance we've already done this and refer you to our earlier article (March 1971 R&T) on improving the performance of Volvos.

The modified Volvos R&T tested are only two examples of the wide range of performance options now available to a Volvo

Volvo sedans lack full instrumentation, so IPD has added a Smiths tachometer, three gauges above the heater controls. A small-diameter Moto Lita replaces the cumbersome Volvo wheel. At right is the fuel-air mixture control for the electronic fuel injection, on the center console.

owner. The list of parts and accessories sold by IPD and Volvo is so extensive and there are so many stages of tuning—from merely improved handling for the road to an all-out road-racing or rally car—that the hardest decision will be deciding where and how to start. For that we suggest you take advantage of the advice and hard-earned knowledge that both IPD and Volvo are ready, willing and able to offer any Volvo enthusiast. ⊙

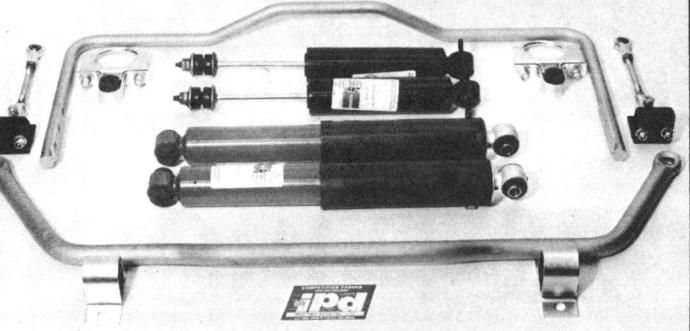

IPD suspension parts: larger front anti-roll bar, rear bar, Heim-joint links for adjusting the rear bar, Bilstein gas-pressure shock absorbers.

Both Volvo and IPD offer several complete camshaft kits which include the camshaft, valve springs, lifters, pushrods and valve-spring retainers. An IPD kit and its detailed information sheet are shown.

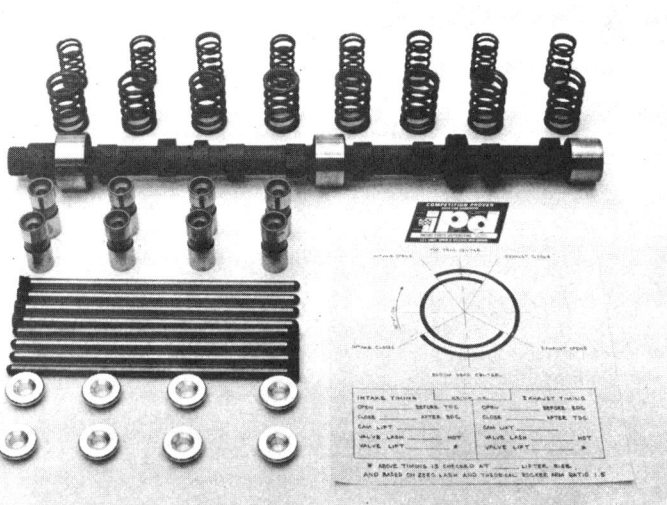

PERFORMANCE EQUIPMENT ON TEST CARS

Volvo Performance Service 142

ENGINE:
Rally camshaft	$102.14
Exhaust header	$72.00

CHASSIS:
Front anti-roll bar	$50.00
Front coil springs	$30.00 set
4.56:1 rear gearset	$87.80
Alloy wheels	$280.00 set

IPD Volvo 142E

ENGINE:
Camshaft	$70.00 exchange
Pushrods & lifters	$48.00
Dual valve springs	$26.00 set
Alloy retainers	$15.60 set
Lightened flywheel	$29.95 exchange
Modifications to cylinder head: port, polish, rework chambers	$195.00 labor only
Alloy valve cover	$33.00
IPD 4-into-1 headers	$96.00 fuel injection
	$89.95 carburetor
Adjustable computer box	$75.00
Engine stabilizer bar	$11.95
Big-bore resonator	$25.00
Clutch	$71.00 cover & disc

CHASSIS:
IPD street coil springs	$48.00 pair, front
	$46.00 pair, rear
1-in. front anti-roll bar	$35.00
¼-in. adjustable rear a-r bar	$59.50
Bilstein shock absorbers	$30.00 each, front
	$34.00 each, rear
American Racing 200-S wheels	$68.50 each, limited supply
4.30:1 rear gearset with Spicer-Thornton Power-Lok differential	$120.00

MISCELLANEOUS:
Wink wide-angle mirror	$19.95
Moto Lita steering wheel	$49.95
Coco mats	$22.50 front & rear
Tachometer	$42.00
Instrument cluster	$86.10
Front spoiler	$39.95
Mallory CD ignition	$75.00

ACCELERATION—BRAKING—HANDLING

	Volvo 142GL	VPS 142	IPD Volvo 142E
Time to speed, sec:			
0-30 mph	4.1	4.0	3.8
0-60 mph	12.7	11.8	10.2
0-80 mph	24.5	23.2	18.3
Time to distance, sec:			
0-¼ mi	18.7	18.3	17.9
Speed at end of ¼ mi, mph	73.0	73.5	79.0
Stopping distances, ft:			
From 60 mph	165	181	157
From 80 mph	297	295	266
Lateral acceleration, g	0.658	0.726	0.785
Speed thru 700-ft slalom, mph	49.4	53.6	54.2

USED CAR R&T CLASSIC

VOLVO 122-S & P 1800

Used sporting cars for people who think

BY THOS L. BRYANT

Y OU PROBABLY DON'T know the names Assar Gabrielsson or Gustaf Larsson and there's really no reason why you should. But, they were the founders of AB Volvo in 1924 and thus the originators of the Swedish car industry. Gabrielsson was a management expert at SKF, a Swedish ball bearing manufacturer, and Larsson was a young engineer with a considerable interest in automobiles. After three years of hard work, they produced their first car in April 1927, a touring car with a 28-bhp 4-cyl engine. Production that first year totaled 297 cars but over the past 50 years, production has risen nearly a thousandfold. Some of the cars from the Gothenburg plant in recent years have been especially noteworthy to the enthusiast and thus form the basis for this "Used Car Classic."

122-S

T HE FIRST Volvos to appear in America arrived in the mid-1950s and were 444 models. Everyone went around commenting on how cute they were and how they resembled, from the rear, a 1946 Ford. Well, the 444 was in production in 1944, so

perhaps this was another case of our ethnocentricity getting in the way of the facts. However, the 444 and later the 544, which appeared in 1958, proved to be popular among enthusiasts and the Swedish car manufacturer was off and running in the U.S.

In 1956, Volvo showed the 120 Series to the press in Sweden and reaction was quite favorable to the new design. In Europe the car would become known as the Amazon, but in the U.S. the moniker was 122-S. It took two years for the 122-S to make the journey to the American market and in the September 1959 issue of R&T there appeared a laudatory road test report on the car: "And a refreshing new car it is, too: pleasant looking, easy (and fun) to drive, economical and durable in the extreme. It is also refreshing to find a company that actually does something to make its product safe for the occupants, and does it without asinine statements that the public won't buy safety."

The R&T report went on to say, "... the newest import is a handsome car in a reserved way, with no evident ostentation or gaudiness." Also, "... a close examination of the car, along with many miles behind the wheel, brought favorable comments from every tester and rider. Design, construction and general quality are obviously excellent, and there is a pervasive feeling of durability."

The early 122-S was powered by a 1586-cc engine which had been used in the 444 since 1956. In the 122-S, the engine, designated the B-16, developed 85 bhp at 5500 rpm and 87 lb-ft torque · at 3500 rpm. In Sweden, the B-16 engine could be purchased with a single carburetor and detuned output of 60 bhp, but the export model had twin SU carburetors. All of the motoring press was enthusiastic about the performance capability of the 122-S, saying the 4-cyl engine was extremely flexible and one of the most free-revving rocker-arm engines around; pointing out, too, that the car's performance was one of its outstanding attractions, especially considering that it was a family 4-seater. The R&T test in 1959 showed a top speed of 92.0 mph and a 0–60 mph time of 16.2 sec. Certainly not breathtaking performance, but

admirable for a 4-door sedan with a 4-cyl engine.

The 122-S was built as a unit body/frame car for safety and durability. Other safety features included optional shoulder belts and such standard items as a padded instrument panel, dished steering wheel and a collapsible steering column. The package tray on the passenger's side was also designed to collapse under impact and the sun visors were thickly padded—all of this long before governmental regulations came into being.

The suspension design of the 122-S was carried over from the 444 and 544: coil springs and tube shocks all around with A-arms and an anti-roll bar in front, while at the rear there were trailing arms and a Panhard rod for lateral location of the live axle. This combination worked very well and we noted in our original test report that "corners can be taken with gusto, though with considerable body roll and squealing of tires . . ." We also noted that the ZF steering was precise and transmitted good road feel to the driver, although it seemed slower than its 3.2 turns lock to lock would indicate.

For 1962, Volvo added a 2-door sedan to the 122-S line and upped the engine displacement to 1780 cc. Power rose from 85 bhp to 90 at 5000 rpm and the torque went from 87 lb-ft to 105 at 4000 rpm. The increased displacement of the B-18 engine was accomplished by an increase in the bore (from 3.13 to 3.31) while the stroke remained the same as in the B-16 engine. At the same time, disc brakes were now standard on the front instead of drums all around as on the earlier models, and a 12-volt electrical system replaced the 6-volt setup. The new B-18 engine was reported by Volvo to have greater fuel economy as a result of increasing the number of crankshaft bearings from three to five,

full machining of all combustion chambers and two additional intake ports resulting in direct induction to each cylinder. Our tests, however, didn't support the factory claims. We reported that the 122-S with the B-16 engine would return 24–27 mpg (R&T, September 1959) while the B-18 engine delivered 21–26 mpg (R&T, May 1962). The larger engine did show an improvement in performance, however, as the acceleration time from 0–60 mph dropped from 16.2 sec to 14.5.

The new engine was also used in the P 1800 which had been introduced just shortly before the 122-S 2-door, so let's take a look at that model.

P 1800

THE P 1800 was designed in 1959 and introduced in late 1961 as a 1962 model. It shared the same suspension and engine with the 122-S but it had 10 more horsepower (100 @ 5500 rpm vs 90 @ 5000) as a result of different carburetion. The body was designed by Frua of Italy and the first couple years' production was built in England with the drivetrain components being shipped over from Sweden. Late in 1964 production was transferred to the Volvo factory in Sweden.

R&T tested the P 1800 in February 1962 and made the following comments about its performance: "All cars of this type that come our way for test get a thorough wringing-out on twisty roads and there the Volvo, despite its weight and soft ride, gave a fine performance. There is a dreadful amount of lean while cornering, but the driver can't feel it inside the car and it doesn't seem to affect the handling. Bends, fast or slow, can be taken with *élan*—just a touch of steadying understeer being present at all times."

We also noted that the P 1800 was not meant for sprinting and that it would take a drubbing in acceleration from less expensive cars, but went on to say, "In doing that for which it was intended, fast steady cruising, the P 1800 is superb and it gave us the impression it would run forever at near maximum speed."

With the move to production of the 2 + 2 coupe in Sweden the car became known as the P 1800S. The basic changes were a revised interior including new seats, less fancy wheel covers and a boost in the bhp figure to 115 at 6000 rpm. Despite the increased horsepower, performance was virtually unchanged (0–60 mph in 13.9 sec for the 1800S vs 13.6 sec for the original P 1800) and it was more or less a case of Volvo continuing with what they considered a basically good car that needed only refinement. By the time of the R&T road test of the 1800S (August 1966) our staff had become less than enthralled with the styling of the car: "... staff opinions on the 1800S styling were generally unenthusiastic, with low marks going to the chromium sweepspear and the semi-finned rear fenders, both cliches of a bygone American era."

In 1969, the 1800 went through another evolutionary step, with the engine jumping from 1800 cc to a full 2 liters with fuel injection. The car was designated the 1800E and there were other important changes as well, including 4-wheel disc brakes, Michelin radial tires, aluminum alloy wheels and a strengthened gearbox. The eggcrate grille was replaced by a simpler and more attractive one composed of horizontal bars, but the basic styling was the same.

The 2-liter, fuel-injected engine gave the 1800E greatly improved performance, lowering the 0–60 mph time to just over 10 sec. The 2-liter engine was also used in the sedans of the day, but only the 1800 received the fuel injection and thus had 12 more bhp and 7 more lb-ft torque than the carbureted Volvo engine at no increase in engine speed. Our comments in the road test of the 1800E (February 1970) went like this: "The engine, noted for its durability rather than refinement—it's neither mechanically smooth nor quiet in the coupe—has good low-speed torque as well as the ability to pull nicely all the way to its 6500-rpm redline, and in overdrive the car will now do an honest 115 mph. The 0–60 mph and ¼-mile times are quite respectable too, putting the 1800E into the same class with such cars as the Alfa 1750, BMW 2500 or Mercedes 280SL. Furthermore, the engine runs. cleanly without any trace of emission-control leanness symptoms and uses very little more fuel than the earlier test car."

Introduction of 1800E in 1969 brought with it a nicely upgraded dash layout.

Volvo 1800S.

Volvo 1800E.

Driving Impressions: Volvo 122-S & P 1800

I was about half way through this project when I looked out the window into the parking lot one day and spotted a good looking 122-S there. "Eureka," I said to myself, "for once I won't have to chase all over southern California searching out a representative car to drive." Joe Bergman is the Assistant Managing Editor of *Sea* Magazine, one of our companion publications, and he owns the 122-S pictured here.

Joe bought the car, a 1968 model, in Indianapolis in the fall of that year. He is the original owner and has racked up more than 104,000 miles on his Volvo. Through the years he has replaced the clutch and universal joints once and installed hydraulic clutch cylinders three times. Other than that, only routine maintenance has been necessary.

I borrowed the 122-S one morning for a drive along Pacific Coast Highway. My first impression was one of solidity: the body, chassis, engine and gearbox all have a solid, unbreakable feel. The steering is slightly heavy by today's standards and that too adds to the overall impression of durability.

The performance characteristics of the 122-S are not startling but it is a fairly responsive car. The braking ability I would call adequate although not superior. Hard cornering produces lots of body roll but once you get used to that and press on, the car actually corners quite well.

There is more than enough leg and head room for the taller driver and Joe's car has the more modern orthopedic seats than those found in earlier versions of the 122. It's a very comfortable sedan, although leg room for the rear seat passengers is somewhat limited with the individual front seats pushed back to their limits. All in all, it is a sports sedan with adequate handling characteristics, ample though not overpowering performance and superior durability.

P 1800 Impressions

Peter Alper is the Advertising Manager for Volvo of America's Western Division and he provided a considerable

Our 1970 test shows we were quite favorably impressed with the new gearbox: "The hefty lever on the new gearbox gives one an impression of unbreakability that is borne out by the gearbox itself; we manhandled the box unmercifully in the acceleration tests and found it capable of taking the fastest slam shifts without a crunch."

In summarizing the 1800E in 1970, we noted that while the car was up to date in performance, handling and braking, the styling, accommodations and use of available space had fallen off the pace of cars at that time, and called on Volvo to design a new model.

Buying A Used Volvo: What to Look For

ALTHOUGH WE are covering rather distinct models in this "Used Car Classic," we are going to make some generalizations in this area of the story and, of course, point out individual quirks or characteristics where necessary. The engines, from the B-16 to the B-18 to the 2-liter B-20, are workhorses. R&T printed an Owner Survey covering the 1800, 122-S and early 144 Volvos in the March 1969 issue and not a single car in that report had required an engine overhaul. We were impressed by that and said in the survey, ". . . we're going to project 110,000 miles as an average life between overhauls in a Volvo." When searching for a good, used Volvo then, chances are the engine will not require major work, but it should be checked over anyway, and preferably by someone who knows Volvos pretty well.

Some of the early 122-S models came with a 2-speed automatic transmission that was, frankly, terrible. In keeping with the tenor of this report, we would omit those cars with an automatic gearbox from consideration in purchasing a used Volvo because of the diminished performance. The 4-speed manual gearboxes found in all of the models have proved nearly indestructible. Having said that, however, we should point out that on nearly all the early 1800 coupes, the overdrive unit shared a common oil reservoir with the gearbox. The Laycock de Normanville overdrive used 30W engine oil for lubrication and many owners never bothered to check the oil level. The result was that they fried the overdrive. So, perhaps the first question to ask in examining an 1800 is, "Does the overdrive work?"

Unlike many of our earlier "Used Car Classic" subjects, Volvos have not been prone to body rust problems. This is not to say that it never occurs, but that it's not especially common. However, we should mention the early 1800 coupes had an annoying tendency to leak water through the cowl vent in front of the windshield. Volvo is quick to point out that this was a problem on those cars that were built in England. Often, this is a result of the deterioration of the rubber molding around the vent hatch or it may be that the drain tubes that allow excess water to run out of the vent opening are clogged.

Our 1969 Owner Survey indicated that many 1800 coupe owners found the instruments rather unreliable, that cooling system problems could occur in all models, that 122-S cars tended to have window winding problems and 10 percent of those responding to the survey reported oil leaks from the differential and gearbox. The survey also showed that 12 percent of the cars had clutch difficulties.

By and large, it would be safe to say that the Volvo models covered in this "Used Car Classic" probably have fewer major problems than most any other car we've reported on in the past. The prospective buyer must exercise caution, of course, to be sure the car he or she wants is in good mechanical condition, but with the Volvos the chances seem to be better than average. Ⓥ

---

amount of expert help in the preparation of this report. In addition, Peter owns a 1963 P 1800 and he brought it by for us to drive. Peter's Volvo is not typical but we thought it would be of interest to discover what can be done to build a street racer.

Alper bought his P 1800 four years ago for $200 and took it home on a trailer. "It was junk," according to Peter, "and I knew it was going to be a total rebuild operation." The first step was installation of a 1973 B-20 (2-liter) engine in place of the B-18. The new engine was given an Iskenderian VV71 camshaft and tubular pushrods, Volvo Competition Service (VCS) light-weight lifters, VCS exhaust

header, intake manifold and dual Solex carburetor set up. Peter had the head redone too but kept the stock ignition system because he feels it's better than anything else he could add. A new 4-speed gearbox and rebuilt overdrive unit along with VCS heavy-duty clutch and pressure plates also went in, as did VCS rally springs at the rear end.

There have been a number of other modifications and changes including a Volvo 140-series sealed cooling system, Bilstein shock absorbers, IPD (specialists in Volvo handling and performance parts, 2762 N.E. Broadway, Portland, Ore. 97232) front and rear anti-roll bars, 7-in. wide American Racing wheels, Pirelli CN36 radials, a custom front air dam and lots more. Peter estimates he's getting 165 bhp from the engine and says the car will do an honest 125 mph.

Driving this P 1800 has almost nothing to do with what you would experience if you went out looking for one to buy, but it is great fun and damn exciting. As you would guess, it has more acceleration than you can normally use and Peter's alterations to the suspension have brought about dramatic changes in the handling. In the final production years of the P 1800, our road test reports were critical of the car's outdated feel: the suspension was too soft, the performance was not crisp, and so on. Alper's modifications show that these problems can be overcome and the car converted into a strong performer. — *Thos L. Bryant*

PHOTO BY JOE RUSZ

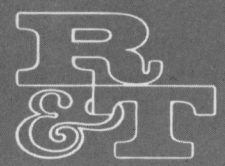

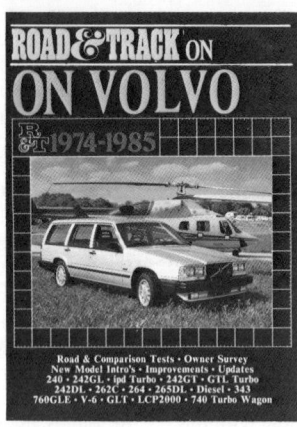

ROAD & TRACK

ON VOLVO 1974-1985